Grasses and Grassland Aspects

Edited by Valentin Missiakô Kindomihou

Published in London, United Kingdom

IntechOpen

Supporting open minds since 2005

Grasses and Grassland Aspects
http://dx.doi.org/10.5772/intechopen.77919
Edited by Valentin Missiakô Kindomihou

Contributors
Sali Aliu, Imer Rusinovci, Dukagjin Zeka, Valentin Missiakô Kindomihou, Kristofor R. Brye, Tyler Durre, Lisa Wood, Edward Gbur, Xue-Feng Ma, Tadele Kumssa, Joshua D. Anderson, Twain J. Butler, Daniel George Neary, Jackson Leonard, Sharon Clay, Alexander Smart, David Clay, Xuhui Zhou, Guiyao Zhou, Lingyan Zhou

First published in London, United Kingdom, 2020 by IntechOpen
IntechOpen is the global imprint of INTECHOPEN LIMITED, registered in England and Wales, registration number: 11086078, 7th floor, 10 Lower Thames Street, London, EC3R 6AF, United Kingdom
Printed in Croatia

British Library Cataloguing-in-Publication Data
A catalogue record for this book is available from the British Library

Additional hard and PDF copies can be obtained from orders@intechopen.com

Grasses and Grassland Aspects
Edited by Valentin Missiakô Kindomihou
p. cm.
Print ISBN 978-1-78984-949-3
Online ISBN 978-1-78984-950-9
eBook (PDF) ISBN 978-1-78985-197-7

For EU product safety concerns:
IN TECH d.o.o., Prolaz Marije Krucifikse Kozulić 3, 51000 Rijeka, Croatia, info@intechopen.com or visit our website at intechopen.com.

Meet the editor

Valentin Missiakô Kindomihou (PhD) is Professor of Animal Production and Agrostology at the School of Animal Production, Science and Technologies, Faculty of Agronomic Sciences, University of Abomey-Calavi (UAC-Benin). He graduated as an agricultural engineer in Animal Production and Technologies, and has an MBA in Environmental Protection from the University of Niamey (Niger), an MSC in Evolutionary Genetics and Plant Ecology and a PhD in Agronomy and Bio-Engineering from the Free University of Brussels (Belgium), and a postdoctoral from Okayama University (Japan). He has 25 years of research and teaching experience, is a consultant for agencies and companies, and a reviewer for scientific journals. He has produced about 130 publications, including one book, 15 book chapters, more than 75 research papers, and 50 communications. His research interests include pastoralism, pasture ecology, biodiversity, management, agrostology, ecophysiology, forage biotechnology (storage, conservation), and organic livestock farming.

Contents

Preface

Grasses and grassland are tools for an integral ecology. Putting this science in the category of well-being is still relevant today. Whether grasses and grassland can meet the present and future challenges requires an update on empirical and scientific achievements and the life quality on earth.

Environmental integration into politics has led to experts and decision-makers facing (1) problem complexity, (2) risk invisibility, and (3) strong future uncertainties. In light of these certainties on which to base decision-maker's actions, research is needed on social challenges regarding environmental crises.

This book project has been initiated by IntechOpen. It is indeed a privilege to have this book published. An array of specialists was invited to contribute seven chapters.

The present volume therefore shows how grasses and grassland can deal with life issues and meet societal expectations: food, environmental, ecological, medicinal, energy, economic, etc. These issues require representation because of environmental changes and their solutions need to be assessed. However, this decision-making process must be undertaken in a context of high complexity and uncertainty.

Grasses and grassland are multifunctional for society, expressing a great diversity of flora and fauna, sustainably managed by grazing, and adapted to the objectives of preserving and restoring diversity, which are ecological and agronomic issues.

By storing carbon and other key nutrients in soils, grasses and grassland provide biological feedstock for developing agricultural and associated human societies feeding humans, supporting domestic livestock production for human use, and using for biofuel production.

Moreover, grasslands are known for developing and testing ecological theories for assessing the impacts of global change, including responses to chronic nitrogen deposition, elevated CO_2 concentrations, and climate change. These include the most threatened ecosystems, such as the North American tall grass prairies. Conserving or restoring these ecosystems will require a strong ecological knowledge base.

In this context, this book aims to provide the reader with a better understanding of the achievements of grasses and grassland, values, and the response to management factors, including grazing, fires, pesticides, fertilization, and associated functional consequences using as models of study: (1) degradation of soil vegetation and ecosystems; (2) pastoral fires; (3) pastoral use of crops and residues; (4) soil–plant–atmosphere continuum; and (5) fodder cropping sustainability.

Varied perspectives on a common interest—the future of grasses and grassland science—reinforce the book's imitative and dominant subject matter. Progress in agronomy, ecology, quality, evaluation, and applications is highlighted with new directions for future research and development being explored.

Apart from the introductory chapter, most chapters are co-written by scientists from various institutions and specialties. Thus, different perspectives on a particular subject are highlighted from one chapter to another throughout the book. In addition, areas where more information is needed and where new research efforts should focus are identified.

Finally, I extended my gratitude to all the authors, sponsors, and funders whose efforts and financial support have helped to make this book project a reality. Special thanks to my mentor Prof. Brice Sinsin (PhD Agronomy), agro-pastoralist, forester and biodiversity conservator, Chair of Tropical Ecology (Plants and Animals), Head of the Laboratory of Applied Ecology of the Faculty of Agronomic Sciences (UAC), and former Rector of the University of Abomey-Calavi (UAC-Benin Republic) for his many investments in building and promoting excellence in tropical science; Roland Holou (PhD.Agronomy), agro-pastoralist and DiasporaEngager, CEO (Georgia, USA); and Ozias Hounkpatin (PhD Agronomy) of the University of Uppsala (Sweden) for their collaboration. The Beninese Association for Pastoralism (ABEPA), International Society for Silicon in Agriculture (ISSAg), and Grassland Society of Southern Africa (GSSA) are also acknowledged as training networks. Bishop Barthelemy Tiando Bona and pastors Arnaud Assogba and Mathias Doukpo of the Assemblies of God Churches offered supportive prayers. My spouse Irma Cale Pascaline Kpakpo-Kindomihou (BSc Accounting) and my child Kabôd Gnonannon Kindomihou offered a patient working atmosphere.

Last but not least, I wish to express my deepest appreciation to the IntechOpen staff, i.e. Iva Simčić, Ivana Spajić, Edi Lipović, Dan Brooker, Sara Uhač, Adrian Assad De Marco, Chris Felder, Alex Lazinica (PhD), Vedran Kordić, and Anke Beck (PhD), CEO, for their timely effort in the publication of this book.

Valentin Missiakô Kindomihou (Ph.D. Agronomy)
University of Abomey-Calavi,
Benin Republic, Western Africa

Chapter 1

Introductory Chapter: Milestones in *Grasses and Grassland* Research

Valentin Missiakô Kindomihou

1. Introduction

In order to understand the significance of changes that have occurred in *Grasses and Grassland* Research and utilization, a short summary of the state of knowledge is required. The purpose of this chapter will be to briefly describe the milestones in *Grasses and Grassland* Research that have occurred. *Grasses and Grassland* are key links in biological resources. They are the sources of many agricultural productions, livestock systems, and environmental issues with positive and recognized impacts on water quality, biodiversity, and landscapes. However, their acreages have been steadily decreasing for many years.

Otherwise, if the livestock structures trend is accused, there is also the lack of knowledge of technicians about the potential of these areas so sensitive to climatic hazards, unproductive, and difficult to be managed. For example, securing fodder systems in organic ruminants farming remain questionable facing these plagues. *Grassland* would interest such systems by making them productive, stable, input-efficient, and environmentally friendly and guarantee good technological performance. In addition, the complexity of understanding the functioning of *Grassland* covers requires synergetic interdisciplinary skills (phytosociology, agronomy, animal technologies, etc.). Optimizing knowledge of *Grasses and Grassland* at scales of production systems should help conceiving self-sufficient, resilient, and sustainable livestock systems, which would meet society's new expectations.

In order to improve knowledge on these issues, this chapter aims to provide scientists, students, technicians, decision-makers, and other development actors with benchmarks for the diagnosis and management of these resources and ecosystem. Topics which were discussed specifically include properties, synthesis, and some applications of *Grasses and Grassland*. Ultimately, what is the contribution of *Grasses and Grassland* to the sustainable well-being of the living?

A "*Property*" is either (a) an original trait, concept, area of research leading to a significant advance in understanding; or (b) a research review acting as a base for further study and development. A "*Synthesis*" is a physical milestone or the production of a substance by the union of chemical elements, groups, or simpler compounds, or by the degradation of a complex compound (www.merriam-webster.com); while "*Application*" is a use to which *Grasses and Grassland* and components are put.

2. Some general points

2.1 What is "Grasses and Grassland"?

Grasses refers to the monocotyledonous green plant, mostly world widespread, rustic over times and circumstances. The Poaceae family as "real *Grasses*" includes

IntechOpen

cereals, forages, and fodders plants from pastures and lawns. The concept appeared in the fifteenth century, derived from the root "grow." *Grasses* design the climatic vegetation in large areas of low rainfall.

Grassland is a community of herbaceous plants mainly *Grasses*, i.e., a grassy area that last at least several years. It is also a forage crop, mainly composed of *Grasses* and legumes, for grazing or mowing.

About 77% of plant species are *Grasses* on *permanent Grassland* exceeding 80% in the spring and only 65% in the fall; the diversity gradient is between 10 and 40 species [1]. Knowing *Grassland* typology may help build sustainable production systems [2].

Indeed, about 18 types of *Grassland* are distinguished [3, 4]:

a. *Seven types of Grassland* occur based on *duration and composition*: (i) *Temporary Grassland* established on monocropping *Grasses* or fodder legumes, by their association for a short or long duration (i.e., 1–3 years versus 3–10 years); (ii) *Artificial Grassland* mainly sown with fodder legumes for 2–5 years, usually in a crop rotation; (iii) *Permanent Grassland*, long-sown (>10 years) with one or various *Grasses* and legumes species; (iv) *Semi-natural Grassland*, i.e., permanent *Grassland* with native *Grasses*, herbaceous, brush, trees, or groves species; (v) Range as wide areas holding native species that are grazed, fenced, or not; (vi) *Natural Grassland* enriched by native *Grasses* for a long time; (vii) *Multi-species Grassland* includes at least three different species from two different families, which ensure more regular production throughout the year and are more resistant to climatic hazards.

b. *Three types of Grassland defined by the objectives*: (viii) *Grazing meadow*; (ix) *Mixed Grassland* for grazing and mowing; and (x) *Hay meadow* for hay or silage in piles or with wrapping.

c. *Four types of Grassland* rise by *the situation and environment* [5], such as: (xi) *Woodland meadows* as part of a dense and meshed network of hedges, i.e., the hedgerows; (xii) *Wet meadows* rich in biodiversity under moderate grazing, regulating watercourses, preventing flooding, and maintaining habitats in open environments; (xiii) *Dry meadows* of the thermophiles and dry hillsides, with low agronomic value and high floristic richness, shelter threatened species; and (xiv) *Pharmaceutical meadows*, i.e., artificial land planted for both soil enriching and medicinal resources provision for livestock.

d. *Four types of Grassland* highlighted on the *ecosystem production and service system perspective*: (xv) *Agricultural Grassland* essentially *temporary Grassland* with at least two species, for stock building (mainly silage) and green feeding, but also including the most fertile permanent mowed *Grasslands*; (xvi) *Environmental Grassland* belong to *permanent Grassland* grazed by dairy heifers and suckling herds, including *Grasslands* mowed by unfertile environments and used in late mowing for ground-dried hay; (xvii) *Meadows "Close to ecological intensification"* including *permanent Grasslands* both grazed (by suckled herds, heifers, and dairy cows) and mowed for hay. Agricultural results are good in all areas: yield, flexibility, quality; and (xviii) *productive meadows*, which are part of the mono-specific, *temporary grassy*, low-*agricultural meadows*, lacking operational flexibility and energy quality, and used for silage, hay, or pasture.

2.2 Some milestones in Grasses and Grassland research

About 60 years of scientific investigations performed in sub-Saharan Africa resulted in a high biodiversity of *Grasses and Grassland*. Previously, 9700 *grass* species with higher biomass production were globally reported. But few are more widely grown to establish *Grassland*. Their adaptation to thermal, mineral, and water stress, resistance to diseases and pests, biomasses and productivities, seed production and nutritional values were studied. Natural and artificial crossings and new genetic techniques improved the species and offer physiological and morphological characteristics including leaves and stems rates, duration of vegetative cycle, sustainability, sexual reproduction, and vegetative and apomictic reproduction. *Grasses* provide raw materials for human and animal nutrition, i.e., food grains and forage. They have reached an advanced stage of development, such in miniaturizing floral pieces and specialization in various environments.

Globally, 150 *Grass* species were well investigated in tropical Africa [6]. This number was completed during the last 30 years (1989–2018). The Laboratory of Applied Ecology of Professor Brice Sinsin from University of Abomey-Calavi (Benin) has described about 100 additional *Grass* species. Two groups of *Grasses* exist based on development cycle duration:

- Annual species grown in rainy season wither and die. Their reproduction requires mature seed, whose formation can be hindered by intense plant exploitation [7].

- Perennial species with roots and lateral buds located at ground level in tropical zone [8], persistent for several consecutive years.

Grasses are well defined based on specific biological, agronomic, and nutritional characters (**Figure 1**), with three types of morphology, including:

- Erect *Grasses* showing one single axis with reduced basal branching, no shelf of tillers and dotted distribution of ground cover [3, 4].

- Bunching *Grasses* with tillers in many clumps, spots, beaches, or large areas, reproduced by seed, resisted to drought and burnings throughout leaf sheaths arrangements that protect the buds and mostly well adapted to intertropical zone [3, 4, 9]. The most common are *Andropogon gayanus, Hyparrhenia rufa, Panicum maximum*, and *Pennisetum purpureum*.

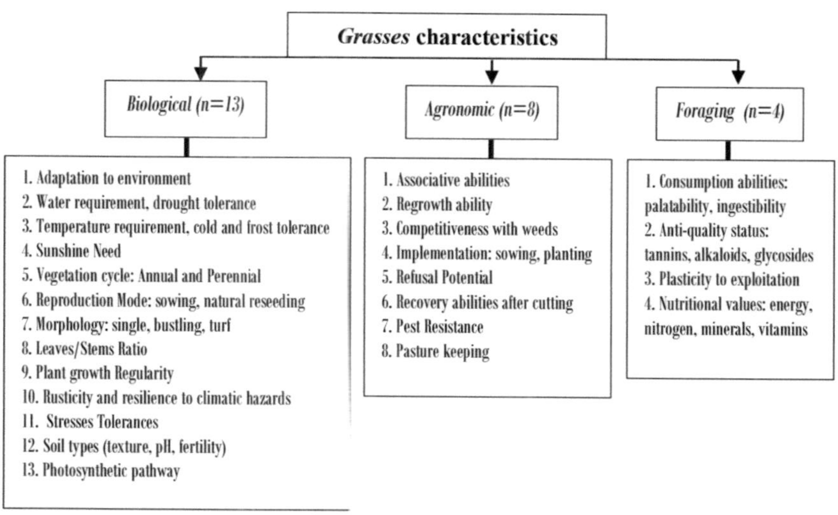

Figure 1.
Grasses *main characters (adapted from [10]).*

- Turfed *Grasses* with rhizomes, allowing horizontal colonization. It forms a dense-felted *Grass*, little fire-resistant, producing little viable seed, and propagates by vegetative pathway [3]. They are the main wetland pasture holding mainly *Brachiaria humidicola, Brachiaria mutica,* and *Brachiaria ruziziensis,* which produces many viable seeds, *Cynodon dactylon, Digitaria decumbens*, and *Stenotaphrum secundatum.*

Grasses, (i) require light to develop, growing in full sun or little shaded; (ii) colonize open fields with high seed production, regenerative capacity, and soil cover rate; (iii) fast-grow with high leaf-to-shoot ratios. The number of degree days to issue a sheet belongs to discriminative indicators.

Forage *Grasses* palatability reflects tissues soft texture, unobtrusive taste, and odor untainted by unpleasant or repulsive substances, low content of toxic substances (tannins, alkaloids, cyanides, and nitrates), high digestible carbohydrate content, nitrogenous matter, easy abilities to be eaten by tearing on the spot, and their abundance. However, these characters are distributed within the same family. Some species are abandoned while others are overgrazed in the same *Grassland*. Foliar nutritional qualities are low in tropics, less digestible because of higher contents in lignin and fibers, higher refusal rate than in temperate environments (i.e., 10–30% versus 5%).

Tropical *Grasses* differ from temperate at the photosynthesis basic energy metabolism. Tropical have a C_4 carbon cycle and Temperate, a C_3 cycle. C_4 *Grasses* store energy for carbon chains production at night from where they grow faster while *Grasses* at C_3 produce carbon chains only in light presence.

These cycles lead to carbohydrates mainly cellulose production. The C_4 cycle is efficient in tropics with less nitrogen than in the C_3 cycle. Species at C_4 optimally develop at higher temperatures than C_3 species (between 30 and 40°C versus 20–25°C) with rapid growth, making maximum use in a short growing season. C_4 *Grasses* produce two to three times more biomass than C_3. Otherwise, they produce more membranes rich in little digestible tissues, with low contents of digestible nitrogen, hence a much less good food value.

Vegetation steps consist of six phases (**Figure 2**).

- *Grasses* can grow in association with other plants based on management practices limiting interspecific competitive effects. Association with legumes offers advantages for soil enrichment in nitrogen mobilized by the legume and

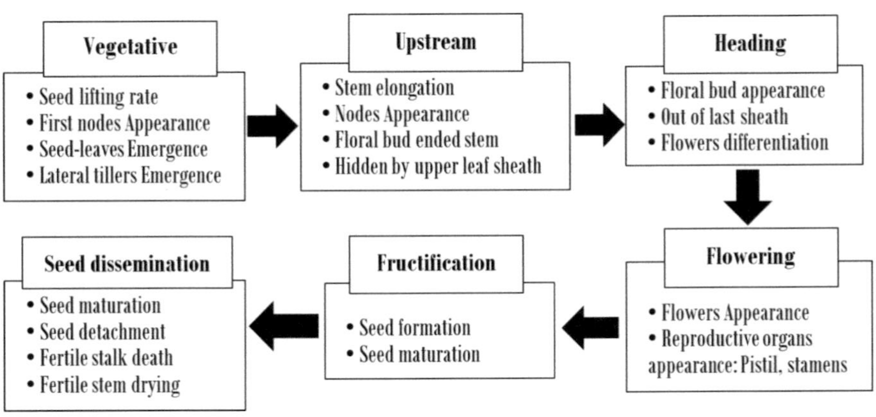

Figure 2.
Grasses *phenological phases (adapted from [10]).*

N good absorption from soil by *Grasses* [11]. Rhizobacteria close to some *Grasses* roots allow atmospheric N recovering, a modality less effective than legumes symbiosis.

- *Grasses* have fasciculate roots from the neck, thatch with nodes, and internodes wearing leaves. These form a sheath with elongated blade between sheath and blade, presence of ligule, small and numerous flowers, each wrapped by two lemmas lower and upper made of two opposite pieces, shells protecting the egg (seed), spikelets, or elemental inflorescence units composed of flowers surrounded by a lower and a upper glumes; inflorescences at top of stem, are composed of spikelets grouped into ears, panicles, digested, or geminated.

3. *Grasses and Grassland'* traits from research

About 78 parameters are listed as indicators for describing *Grasses and Grassland* (**Table 1**). However, molecular aspects need more attention as well as their warming and adaptive responses.

No	Indicators	No	Indicators
1	Ability to compete with weeds	40	Minimum temperature of growth
2	Ability to spread naturally	41	Natural habitat
3	Ability to zerograzing	42	Number of seed per kg
4	Altitude range	43	Nutraceutical properties
5	Allelopathy properties: phytomolecules and compound released	44	Nutritional composition range with climate dependence (proteins, lipids, fibers, ash)
6	Amino acid contents (lysine, glutamic acid, leucine, alanine, proline, aspartic acid, valine, phenylalanine, serine, isoleucine, arginine, threonine, glycine, tyrosine, histidine, methionine, cysteine, tryptophan, phytosterols, polyconasols)	45	Optimum temperature for growth
7	Animal preference	46	Palatability
8	Animal production	47	Pests
9	Anti-quality elements (Tannins, SiO_2)	48	Phenolic acids and phenols contents
10	Bioactive phytochemicals (carbohydrates, proteins, alkaloids, flavonoids, tannins, phenols, saponins, glycosides, steroids, terpenoids)	49	Physicochemical characteristics of grains
11	Chemical analysis and digestibility	50	Phytochemicals major families (β-glucan, phenolic acids, flavonoids, lignans, tocols, phytosterols, folate)
12	Common names	51	Pigments or coloring agents
13	Compatibility with grasses and legumes	52	Quality index
14	Cultivars	53	Rainfall requirements
15	Description	54	Refusal rate
16	Digestibility	55	Response to defoliation
17	Diseases	56	Response to fire
18	Distribution	57	Response to photoperiod

No	Indicators	No	Indicators
19	Dormancy	58	Season of growth
20	DRIS norms, i.e., Diagnosis and Recommendation Integrated System (N/P, N/Ca, N/Zn, K/P, P/Ca, S/P, Zn/P, K/Ca, S/K, K/Zn, S/Zn, Ca/Zn, S/Ca).	59	Seed or grain color, shape, size and anatomical elements
21	Drought tolerance	60	Seed production and harvesting
22	Dry- and green- matter yields	61	Seed treatment before planting
23	Economics	62	Seed vigor
24	Equilibrium moisture of hay	63	Seed yield
25	Fat and fatty acids contents	64	Soil requirements
26	Fertilizer requirements	65	Source of variation levels of compounds released and effects
27	Frost tolerance	66	Sowing depth and cover
28	Further reading	67	Sowing methods
29	Genetics and reproduction	68	Sowing time and rate
30	Grain composition (energy, DM, lipid, fibers, ADF, seed, proteins, ash)	69	Suitability for hay and silage
31	Grazing management	70	Therapeutic properties
32	Growing systems	71	Tolerance to flooding
33	Land preparation for establishment	72	Tolerance to herbicides
34	Latitudinal limits	73	Tolerance to salinity
35	Main attributes	74	Toxicity
36	Main deficiencies	75	Value as standover or deferred feed
37	Maximum temperature, low temperature	76	Value for erosion control
38	Minerals content (K, P, Mg, Fe, Ca, Na, Zn, Cu, Mn, Se)	77	Vigor of growth and growth rhythm
39	Minimum germination and quality required for commercial sale	78	Vitamins contents (Niacin, Vita E, alpha tocopherol, Vit BE, pantothenic acid, thianine, riboflavin, folate, fat, fatty acid)

Adapted from [3, 6–8, 10–14].

Table 1.
Alphabetical fields of information on Grasses *species.*

4. *Grasses and Grassland diversity*

4.1 Cereals

A cereal is a plant grown mainly for its seeds (grains, fruits, and caryopses) used as food and fodder, being consumed in flour or grain form. "*Cereal*" refers to *Grasses* grains. About 43 cereals from nine main tribes have been documented including 17 Paniceae, 16 Triticeae, 2 Cyperaceae, 2 Oryzoideae, 2 Poeae, 1 Aveneae, 1 Chloridoideae, 1 Coccinea, and 1 Eragrostideae. The most world widely grown are rice, maize, wheat, barley, and sorghum, respectively. Otherwise, about four pseudo-cereals are also worldwide recognized but not yet well scientifically supported: buckwheat (*Fagopyrum esculentum,* Polygonaceae), quinoa (*Chenopodium quinoa,* Chenopodiaceae), amaranth (*Amaranthus* spp.,

Chenopodiaceae), and sesame (*Sesamum indicum*, Pedaliaceae). Among the several cereals that are used in feeding human over the world, only the high-growth and high-carrying capacity cereals prevent forage shortages in winter. They can also help controlling weeds. Oats are more forage than wheat, barley, rye, and triticale. Potential forage biomass depends on crops, varieties, pathology resistance, and seeding time.

4.2 Native Grasses and Grassland

As for Asia, Europe, America, and Australia *Grasses and Grassland*, African landscape offers about six types of wide *Grasses and Grassland*. **Table 2** shows the values, productivities, and demand for equivalent land (i.e., the opposite of carrying capacities) of *Grassland* highlighting the main dominant *Grass* species over the normal periods, i.e., 30 years.

The southern Sahara reflects a dynamic presence of six *Grassland* groups (**Table 2**, [15]) burst into 50 *Grasslands,* including 6 Sudanese, 12 Guineo-Sudanese, 9 Sudano-Guinean, 6 Guinea-Sudanese and Sudano-Guinean transitional, 9 Northern Sudanese, and 8 Sudano-Sahelian. Apart from the most productive *Grasslands*, which are artificial, the most productive native is *Andropogon gayanus* and

Ecozones	Types of *Grassland* and main *Grasses* species	PV (%)	t DM/ha	ELD (ha/TLU)
Sudanese (S)	*Aspilia paludosa* and *Anadelphia afzeliana* (SP1)	51.5	1.08–9.79	2.11–0.23
	Hyparrhenia involucrata and *Andropogon pseudapricus* (SP2)	37.3	1.21–10.1	1.89–0.23
	Loudetia flavidae (SP3)	23.9	0.38–5.21	6.00–0.44
	Loxodera ledermanii (SP4)	36.8	0.30–6.91	7.60–0.33
	Pennisetum unisetum (SP5)	46.3	0.25–15.76	9.13–0.14
	Setaria longiseta and *Sporobolus pyramidalis* (SP6)	21.4	0.22–5.58	10.37–0.41
Guineo-Sudanese (GS)	*Andropogon chinensis* (GSP1)	65.82	5.00–5.73	0.46–0.40
	Andropogon gayanus and *Hyparrhenia involucrata* (GSP2)	74.27	5.70–6.86	0.40–0.33
	Andropogon schirensis (GSP3)	50.49	5.40–7.12	0.42–0.32
	Andropogon tectorum (GSP4)	35.04	6.06–9.40	0.38–0.24
	Andropogon macrophyllus (artificial) (GSP5)	64.40	9.73–12.46	0.23–0.18
	Brachiaria ruziziensis (artificial) (GSP6)	63.17	6.90–9.15	0.33–0.25
	Ctenium newtonii (GSP7)	35.04	6.33–10.8	0.36–0.21
	Heteropogon contortus (GSP8)	35.04	6.94–11.4	0.33–0.20
	Hyparrhenia smithiana (GSP9)	46.89	5.63–7.10	0.41–0.32
	Imperata cylindrica (GSP10)	22.82	0.29–5.12	7.87–0.45
	Panicum maximum C1 (artificial) (GSP11)	84.28	10.30–13.51	0.22–0.17
	Sporobolus pyramidalis (GSP12)	56.02	6.53–8.80	0.35–0.26

Ecozones	Types of *Grassland* and main *Grasses* species	PV (%)	t DM/ha	ELD (ha/TLU)
Sudano-Guinean (SG)	*Andropogon schirensis* and *Andropogon gayanus* (SGP1)	45.74	5.26–8.50	0.43–0.27
	Andropogon schinensis and *Andropogon chirensis* (SGP2)	53	4.70–8.80	0.49–0.26
	Brachiaria falcifera (SGP3)	36.88	3.64–7.50	0.63–0.30
	Pennisetum polystachion and Andropogon gayanus (SGP4)	28	3.93–7.20	0.58–0.32
	Pennisetum polystachion and *Hyparrhenia involucrata* (SGP5)	49.33	7.86–9.80	0.29–0.23
	Pennisetum unisetum and *Rottboellia cochinchinensis* (SGP6)	36	3.24–6.80	0.70–0.37
	Sorghastrum bipennatum and *Schizachyrium sanguineum* (SGP7)	50	4.90–7.90	0.47–0.29
	Sporobolus pyramidalis and *Hyparrhenia involucrata* (SGP8)	32.15	5.70–8.40	0.40–0.27
	Schizachyrium sanguineum and *Hyparrhenia rufa* (SGP9)	45.53	4.89–7.01	0.47–0.33
Transition Guineo-Sudanese/ Sudano-Guinean (TGS/SG)	*Andropogon schirensis* and *Hyparrhenia subplumosa* (TP1)	45.8	7.01–10.2	0.33–0.22
	Andropogon tectorum and *Chromolaena odorata* (TP2)	47.6	7.47–10.9	0.31–0.21
	Brachiaria falcifera (TP3)	44.8	5.57–8.40	0.41–0.27
	Hyptis suaveolens and *Hyparrhenia suplumosa* (TP4)	38.2	2.03–5.20	1.12–0.44
	Pennisetum polystachion and *Securinega virosa* (TP5)	48.15	8.25–11.2	0.28–0.20
	Sporobolus pyramidalis and *Hyparrhenia subplumosa* (TP6)	33.4	7.00–11.4	0.33–0.20
Northern Sudanese (NS)	*Andropogon gayanus* (NSP1)	46.5	4.40–6.10	0.52–0.37
	Andropogon gayanus and *Hyparrhenia involucrata* (NSP2)	32.71	7.53–7.98	0.30–0.29
	Andropogon gayanus and *Schizachyrium sanguineum* (NSP3)	52.5	4.80–5.02	0.48–0.45
	Andropogon pseudapricus and *Pennisetum polystachion* (NSP4)	16.77	5.36–5.87	0.43–0.39
	Andropogon pseudapricus and *Tephrosia pedicellata* (NSP5)	14.89	5.64–5.94	0.40–0.38
	Hyparrhenia involucrata (NSP6)	36.2	3.95–5.90	0.58–0.39
	Hyparrhenia involucrata and *Andropogon gayanus* (NSP7)	38	4.80–6.30	0.48–0.36
	Loxodera ledermannii (NSP8)	42.69	6.58–7.10	0.35–0.32
	Pennisetum pedicellatum (NSP9)	23.9	2.60–5.50	0.88–0.41
Sudano-Sahelian (SS)	*Andropogon pseudapricus* (SSP1)	26.28	2.83–3.21	0.81–0.71
	Andropogon pseudapricus and *Panicum pansum* (SSP2)	31.56	4.21–5.12	0.54–0.45
	Diheteropogon amplectens (SSP3)	28.43	3.45–4.10	0.66–0.56

Ecozones	Types of *Grassland* and main *Grasses* species	PV (%)	t DM/ha	ELD (ha/TLU)
	Panicum subalbidum (SSP4)	31.72	3.35–4.08	0.68–0.56
	Pennisetum pedicellatum of savannahs (SSP5)	35.06	2.42–3.45	0.94–0.66
	Pennisetum pedicellatum of fallows (SSP6)	34.05	4.12–4.76	0.55–0.48
	Schoenefeldia gracilis (SSP7)	14.90	3.71–4.01	0.61–0.57
	Vetiveria nigritana and *Oryza longistaminata* (SSP8)	41.44	6.74–7.24	0.34–0.32

Table 2.
Types of tropical pastures, grazing values, productivities, carrying capacities (1988–2018).

Hyparrhenia involucrata from Guineo-sudanese and the lowest is *Setaria longiseta* and *Sporobolus pyramidalis* from Sudanese zone.

5. Grasses and Grassland properties

5.1 Grasses and Grassland food and fodder properties

5.1.1 Native forage grasses

Temperate wild or permanent and extensive *Grassland* can hold up to 100 plant species per hectare. More than 120 native tropical forage *Grasses* in Africa have been studied and found to less contribute to productivities of respective communities. Special attention has been given to their ecology for better production and rational use [3, 4, 12, 16–19].

5.1.2 Introduced fodder grasses

The temperate *Grassland* cultivated or more intensive breeding include the following species: Brome (*Bromus secalinus*), Dactyle (*Dactylis glomerata*), *Festulolium* (more than 40 cultivars), High Fescue (*Festuca arundinacea*), Meadow Fescue (*Festuca pratensis*), Meadow Timothy (*Phleum pratense*), meadow Pâturin (*Poa pratensis*), English Ray-grass (*Lolium perenne*), Italian Ray-grass (*Lolium multiflorum* Lam.), Hybrid Ray-grass (*Lolium* × *hybridum*), legumes species such as Cornish lotier (*Lotus corniculatus*), cultivated Luzerne (*Medicago sativa*), cultivated Sainfoin (*Onobrychis viciifolia*), White Clover (*Trifolium repens*), Alexandria Clover (*Trifolium alexandrinum* L.), Micheli Clover (*Trifolium michelianum* Savi.), Hybrid Clover (*Trifolium hybridum*), Incarnate Clover (*Trifolium incarnatum* L.), and Purple Clover (*Trifolium pratense*).

About 30 exotic *Grasses* were introduced from 1987 to 2015 in Southern Sahara, with 17 mainly fodder [7 *Panicum* species (6 varieties of *P. maximum* and *P. coloratum*) and 5 species of *Brachiaria*] [15, 20]. These samples in seagrass beds and on experimental plots are available in the National Institute of Agricultural Research (INRAB), the Faculty of Agronomic Sciences, other research institutions such as ILRI/ILCA (Kenya), FAO (Rome/Italy), CIAT (Colombia), IDESSA (Bouake/Côte d'Ivoire), Democratic Republic of Congo and ILCA (Nigeria).

Panicum maximum C_1 and *Pennisetum purpureum* are adopted and grown on seedling in density ranging from 30 cm × 30 cm to 40 cm × 40 cm. Several introduced species such as cereals, forage, ornamental and medicinal plants, have already become part of local flora like *Bambusa vulgaris*.

Grasses are for several uses: food, fodder, industrial, medicinal, etc. Smaller local cereals are *Digitaria exilis, Oryza glaberrima, Eleusine coracana*, and *Digitaria iburua. Eleusine coracana*, quite recent expansion is mostly grown in mountains. *Digitaria* spp. are famine consumed, mainly *D. exilis* in West Africa, *D. debilis* and *D. iburua* in East Africa (Cameroon). Industrial cereals are rice (*Oryza sativa*) grown in plains and valleys. Wheat (*Triticum aestivum*) is grown on plates in batch production. *Oryza longistaminata* and *O. barthii* are not yet cultivated. *Panicum laetum* and *Cenchrus biflorus* are eaten locally as well as *Brachiaria xantholeuca, Dactyloctenium aegyptium, Echinochloa colona, E. pyramidalis, E. stagnina, Setaria pumila*, and *Sorghum arundinaceum*. Sugar cane (*Saccharum officinarum*) is grown in rural gardens from the south of West Africa (Benin, Cameroon, Sierra Leone, Ghana, Nigeria, Togo, and Guinea) and mainly used for sucking. *Sorghum*, which is not a new crop, is a regional cereal, subjected to large international commercial transactions. It is one of the most important cereals grown especially in arid and semi-arid land ecosystems. Holding multiple-use properties, all *Grasses* are browsed by livestock, mostly at juvenile stages and recent is *Grasses and Grassland* developments. *Grassland* improvement by livestock farming and forage stations is based on species such as *Brachiaria ruziziensis*, *Cenchrus ciliaris*, *Panicum* varieties, *Pennisetum clandestinum*, *Pennisetum purpureum*, and *Tripsacum laxum*. *Pennisetum clandestinum* is found on some mountain *Grassland*. *Ischaemum timorense* and *I. indicum* belong to introduced species.

A cultivated forage species is selected based on the following criteria:

1. Ability to produce good seeds, spread by runners, rhizomes, or stem cuttings;
2. Being vigorous, high-yielding, palatable and nutritious, leafy, good foliar quality and late flowering;
3. Resistant under intensive grazing;
4. Ability to survive a dry season, provide pasture for a good part of the dry season;
5. Intended for temporary use, being capable of being eradicated with relative ease.

5.1.3 Cereals

Cereals belong to *Grasses* family. Following are some of their feeding properties:

A natural fuel: Cereals hold 70–80% carbohydrates especially starch. They are complex or slow carbohydrates, gradually intestine absorbed, and diffusing energy over time, unlike simple sugars. In its complete form, their shell and fibers slow down this absorption even further. The glycemic index (GI) measures a food ability to increase blood sugar levels within 2 h of ingestion. The higher this GI is, the faster the food is assimilated, and the sooner the feeling of hunger can manifest itself. Whole grains have a low GI, which increases when grains are processed (ground, mixed, blown, extruded ...) because their sugars are then more quickly and easily assimilated.

Vegetable bricks: Cereals are *well* provided with protein a little less than meat (7–14%), the richest being wheat and oats). Optimizing them require combination with legumes (chickpeas, lentils, soya ...), one providing each other with the essential amino acid that they lack to cover human needs (cereals are deficient in lysine, well represented in legumes, which are lacking methionine). Hence, traditional associations are corn + red beans in Mexico, rice + lentils in India, wheat semolina + chickpeas in couscous, and rice + peas in Cantonese rice.

Smooth transit: Unrefined cereals *are* naturally rich in fibers, which, when indigestible, increases stool volume and promotes stool elimination. These fibers also contribute to satiety through a mechanical effect on the stomach.

Mineral wealth: Cereals contain *good* quantities of vitamins of the B group, beneficial for the nervous system, but also vitamin E, antioxidant and many minerals, with mainly magnesium and calcium. **Table 3** highlights some main advantages.

An important cereal food property is Gluten availability. Gluten or prolamin is a protein fraction. Cereals mainly contain it, especially wheat, oat, and barley flour while rice flour is almost exempt from it. Cereals hold sugar (Starch) and protein

Cereals	Characteristics	Use
Barley	Low energy and fat Rich in fibers, Vitamin B_{12}, Gluten	Interesting for vegans Beaded, crushed, flakes, flour
Corn	Sweet taste High glycemic index, energetics	Grains, starch, flakes, flour, syrup Thyroid moderator
Millet	Rich in iron, high vitamin A and phosphorus	Anti-asthenic, invigorating, stimulant infection control, cooked, sprouted
Oats	Richest in proteins (14.2 g), lipids, Ca, Cu Most energetic Gluten light Digestive, diuretic, tonic, hypoglycemic Lower bad cholesterol Thyroid gland stimulators	Flakes + soy milk, soup, porridge Bread, pancakes, Muesli cakes
Rice	Richest in carbohydrates Low in proteins (about 7 g) and lipids Silicon, Vitamins B, fibers	Risottos, pilaf, entremets Flakes, flour
Rye	High proteins (about 12 g) Rich in K and B vitamins Less "dirty" than wheat Contains gluten	Bread, whole, crushed, flakes, flour Soft laxative and depurative Blood thinner
Spelt (primitive wheat)	8 essential amino acids 11–16% of proteins, high Mg Highly digestible Less gluten than wheat	Hypotoxic diet Anti-stress Sensitive stomach and intestines Flour, soups, cakes, coffee substitute
Wheat	Rich in Mg, vitamins B1, PP, E Gliadin	Good for nervous balance Flakes, flour, bread, syrup

Table 3.
Cereals feeding properties.

(Gluten). Gluten contains glutellins and prolamins. There are different types of prolamins and glutellins: wheat, for example, is composed of glutenin (glutin side) and gliadin (prolamin side). It is these combined elements that give elasticity to the bread dough and allow the air bubbles to be enclosed. The higher the prolamin content, the higher the leaven on bread. Several cereals are used in feeding human over the world.

Grasses contain toxic prolamins such as rye secalin, orange hordein, corn zenin, oat avenin, or wheat prolamin (gliadin), which are well known for toxicity. For this reason, these cereals are generally banned from the diet of people more sensitive to gluten toxicity. All cereal *Grasses* contain gluten (**Table 4**). However, prolamin levels are sometimes so low that prolamin intolerants may consume some of them.

Wheat, Kamut, and Spelt are the richest in gluten while rice contains insignificant amounts, as well as buckwheat. These cereals could perfectly fit intolerant persons as gluten-free flours in shops and supermarkets.

Fonio (*Digitaria exile* and *D. iburua*) as the oldest cereal of ethical minorities in arid sub-Saharan Africa is of nutritional properties as the richest in Mg, Ca, Fe, and Zn, which contributes to properly functioning of the immune system. It is gluten-free and contains twice as many amino acids as other cereals. It can be consumed by coeliac disease victims or wheat allergy, as it contains less protein than others and similar to white rice in composition. No major scientific studies have looked specifically at Fonio. However, being considered as a whole grain, we know about their positive impacts on the risk of cardiovascular disease, type 2 diabetes, constipation, overweight, cancers including colorectal cancer.

5.2 Grasses and Grassland anti-erosive properties

Anti-erosive practices favor mounts terraces and foothills covering massifs. Terraces create fields on slopes and among blocks. Millet stalks retain elements from disintegration in place of granite rock slabs of walls sealed with gravel.

Rank	Cereal	Scientific name	Gluten type (prolamin)	Prolamin content (%)
1	Wheat	*Triticum aestivum L. subsp. spelta* (L.) Thell.	Alpha gliadin	69
2	Spelt	*Triticum spelta* L.	Alpha gliadin	69
3	Kamut	*Triticum turanicum* D. Love	Alpha gliadin	69
4	Corn	*Zea mays L.*	Zenin	55
5	Sorghum	*Sorghum bicolor* (L.) Moench	Cafirin	52
6	Barley	*Hordeum vulgare L.*	Hordenin	46–52
7	Rye	*Secale cereale L.*	Secalin	30–50
8	Millet	*Pennisetum glaucum* (L.) R.Br.	Panicin	40
9	Oat	*Avena sativa L.*	Avenin	20–30
10	Tef	*Eragrotis abyssinica* (Jacq.) Link	Not identified	12
11	Fonio	*Digitaria exilis* (Kippist) Stapf *Digitaria iburua Stapf*	Not identified	<10
12	Rice	*Oryza sativa L.*	Oryznin	5

Table 4.
Classification of food grasses *according to gluten content.*

Anti-erosive practices	*Grasses* species used	Organs used	Localities
Terraces	*Adiantum philippense* L. (Ferns) *Cynodon dactylon (L.) Persoon* *Digitaria argillacea* (Hitchcock and Chase) *Rottboellia cochinchinensis* (Lour.) W. Clayton	Stoloniferous roots	Mountains
Rock slabs	*Pennisetum typhoides* (Burm.) Stapf. and Hubb.	Stems	Walls
Improved fallow lands	*Setaria pumila* (Poir.) Roem. and Schult., 1817 *Setaria sphacelata* (Schumach.) Stapf and C.E.Hubb.	Roots, stolons	Uplands
Plant defensive systems	*Sorghum caudatum* (Hack.) Stapf. *Sorghum durra* (Forssk.) Trab.	Stems, stolons	Plains and foothills
Pseudo-terraces Restored halomorphic floors	*Sorghum caudatum* (Lithosols Sorghums) *Sorghum durra* (Transplanted Sorghums)	Stems	Mountains
Bunds gridding	Off- seasoned Sorghums	Stems	Uplands

Table 5.
Anti-erosive grasses *(adapted from IRD1995–1996).*

Gradually, these elements rise until an arena beach becomes a field. Bundles of millet stalks and certain *Grasses* (**Table 5**) are arranged to also retain fine elements [21].

Terraces are maintained by particular crops in the blocks interstices, including *Stoloniferous Grasses* such as *Cynodon dactylon* and *Adiantum philipensis,* i.e., a small fern. *Rottboellia cochinchinensis* and *Digitaria argillacea* are cutting and offer 2–3 cuts for cattle feeding in rainy seasons.

Facing foothills and plains land insecurity, new intensive and elaborate anti-erosive practices help to recover hard land (halomorphic soils), which have often been abandoned as a village cattle parking lot. Farmers dig holes 1.5–2 m apart and 35–40 cm in diameter with as much depth in the compacted horizon. They spread all around the manure, fill the holes in the second or third year, and cultivate a first time with the plow, early *sorghum*. Seasoned *sorghums* are grown annually. At this stage, a grid of bunds of 30 cm height is mounted for better water contention. Planting *Setaria pumila* and *Setaria sphacelata* improve fallow lands [22]. Halomorphous soils restoration occurs between 4 and 6 years. In such context, clusters of rocks, as well as halomorphic areas deemed uncultivated, can give fields. The anti-erosive aspect serves essential crops, *Sorghums* of Lithosols Mountain and transplanted *Sorghums*, all of which are free from fallow.

5.3 Grasses and Grassland ornamental properties

Grasses from wetlands mostly grew for ornamental purposes in dryland irrigation systems (**Table 6**). *Axonopus compressus*, *Cynodon dactylon*, and *Paspalum conjugatum* grew in urban and peri-urban areas, particularly in South Africa (Cape Town, Johannesburg, Pretoria), Benin (Cotonou, Ouidah, Parakou, Porto-Novo), Cameroon (Ngaoundé, Bertoua, Yaounde, Douala), Sierra Leone (Freetown, Njala), Liberia (Monrovia), Senegal (Dakar, Saint Louis, Ziguinchor), Congo (Kisangani, Kinshasa, Brazzaville), Ethiopia (Addis Ababa), Ghana (Cape Coast, Tema, Kumasi, Accra), Nigeria (Abuja, Abeokuta, Shagamu, Idjebu-Ode, Lagos, Ibadan, Ogbomosho, Oyo, Iseyin, Makurdi, Awka, Kaduna, Maiduguri, Yenagoa, Asaba,

Species	Locations	Utilizations	Ecoregions
Axonopus compressus (Sw.) P. Beauv. *Chrysopogon aciculatus* (Retz.) Trin. *Paspalum conjugatum* P.J. Bergius *Vetiveria nigritana* (Benth.) Stapf	Cities bordering African marine Coast	Ornamental, fodder and, medicinal	Tropical wet and dry
Axonopus compressus (Sw.) P. Beauv. *Cynodon dactylon (L.) Persoon* *Paspalum conjugatum* P.J. Bergius	Cosmopilites	Ornamental, fodder and medicinal	Tropical dry and wet
Chrysopogon aciculatus (Retz.) Trin.	Airports	—	Cosmopolites
Cymbopogon citratus (D.C) Stapf.	Home gardens	Ornamental and medicinal	Tropical wet, dry
Oxythenantera abyssinica A. Rich. Munro	Botanical Gardens, Southern Cities (Wetlands)	Fences and medicinal	Tropical wet, dry Temperate
Phyllostachys aurea Koï	Gardens of northern cities (drylands)	Fences and handcrafts	Tropical wet and dry
Polytrias diversiflora (Steud.) Nash *Stenotaphrum secundatum* (Walter) Kuntze	Botanical Gardens	Ornamental and medicinal	Tropical, temperate

Table 6.
Main ornamental grasses.

Calabar, Enugu, Akure, Port Harcourt, Bonny, Ahoada, South and North Sokoto, Lafia, Kano, Owerri, Minna, Ilorin, Gombe, Ado Ekiti, Dutse, Katsina, Benin-City, Abakaliki, Lokoja, Osogbo, Jos, Jalingo, Donga, Damaturu and Gusau), Morocco (Casablanca, Tanger, Marrackech, Raba). In addition, two cosmopolitan seal (*Axonopus compressus* and *Paspalum conjugatum*) with *Chrysopogon aciculatus* mark lawns of African, Asian, European, American and Australian coastal cities. *C. aciculatus* outcrops gardens of East and West African airports.

African landscapes are dominated by bamboo (*Phyllostachys aurea*) while those of northern cities (Europe, America, Australia, and Asia) are distinguished by cultivation of *Oxythenantera abyssinica*. *Cymbopogon citratus* is dominant in ornamental and medicinal gardens. *Stenotaphrum secundatum* marks lawns of East African marine coasts, and *Polytrias diversiflora*, Municipal and University botanical gardens. Linear foliage and flowers in form of ears sometimes feathery appear in autumn and bring a graceful and light touch. Near a resting place or water garden, the rustling and swaying of leaves and ears of corn bring into Morpheus' arms. Moreover, *Miscanthus* (*Miscanthus*), Panic (*Panicum*) and *Calamagrostis* (*Calamagrostis*) are other values.

5.4 Grasses and Grassland aromatic properties

5.4.1 Aromatic properties

Several aromatic species are grown mostly for oils i.e. lemongrass *Cymbopogon citratus* and *Melinis minutiflora* grown in African gardens (**Table 7**). *Vetiveria*

Grass species	Aromes	Utilizations	References
Cymbopogon citratus (D.C) Stapf. *Cymbopogon densiflorus (Steud.) Stapf* *Cymbopogon schoenanthus* (L.) Spreng., 1815	Essential oil Citral	Burnt panicles with odoriferous fumes for ceremonials, anti-amoebic, antibacterial, antidiarrheal, antifilarial, antifungal, antioxidants, antimalarial, hypoglycemic, anti-inflammatory, antimutagenicity, antimycobacterial, neurobehaviorial	[23, 24]
Melinis minutiflora P. Beauv.	Cumin aroma, essential oil	Pesticids, acaricids, ovycids	[25]
Vetiveria zizanoides (L.) Nash	Essential oil	Tenacious smell, perfumery, cosmetics, strong, pleasant and lasting notes from flavors, insecticides, herbicides, anti-microbial, antioxidant	[26]
Miscanthus spp. *Miscanthus sacchariflorus* (Maxim.) Franch	Essential oil	Bio-energy, biofuel, bio-crude oil, diesel	[27]
Arundo donax L. *A. phragmites L.*	Oils, lignin, alkaloids	Biogas, construction, medicinal	[28]
Nardus stricta L., 1753 *Anthoxantum odoratum* L., 1753	Coumarins	Medicinal, culinary, cosmetic and economic *properties*, cultivation	[29]
Bromus catharticus Vahl, 1791	Aroma	Purgative, spines, burrs, toxics, pest host	[30, 31]

Table 7.
Some aromatic grasses, *aromes and usages.*

zizanioides, *Cymbopogon schoenanthus* and *C. densiflorus*, cultivated in Cameroon's Adamaoua region, offer essential oils which are widely used in crop protection, storages, veterinary and human medicines.

5.4.2 Grasses and Grassland essential oils composition

Ninety-five volatile compounds, representing 75.4% of the total area, were identified in essential oils from *Grasses and Grassland* plants, with each of remaining peaks accounting for less than 0.01%. Terpenoid family is mostly abundant with 14 monoterpenes, 24 monoterpene derivatives, 18 sesquiterpenes, and 11 sesquiterpene derivatives, together accounting for 61.1% of total peak area. Besides, were seven benzenic compounds accounting for 12%. Benzenic compounds dill apiole and carvacrol were mostly abundant in essential oil after sesquiterpene germacrene D. The other compounds (1 ketone, 6 aldehydes, 4 alcohols, 3 esters, and 7 alkanes) accounted for 2.3%. Essential oil from *Grasses and Grassland* contained the usual terpenes [32–34].

5.4.3 Grasses and Grassland *medicinal properties*

All organs from *Grasses* are found with specific medicinal virtues for human and animal well-being (**Table 8**).

Grass seeds generally exert a particular action on the nervous system, which result in dizziness, and a body tremor as the case for ryegrass and *Festuca*

Scientific name	Common name	Part used	Dosage form	Medicinal properties	References
Andropogon spp	True lemon grass	Leaf, root, stem	Oil, paste	Anti-arthritic	[35, 36]
Cenchrus ciliaris L.	Buffel grass	Aerial and root part	Oil	Anti-cancer	[37]
Cymbopogon spp	Melissa grass, Geranium grass	Leaves	Oil	Anti-amoebic, antibacterial, anticancer, antidiarrheal, anti-filarial, antifungal and anti-inflammatory, antimalarial, anti-mutagenicity, antioxidants, anti-arthritic, anti-mycobacterial, hypoglycemic, neurobehavioral	[24, 38]
Cynodon dactylon Pers.	Bahama grass	Plant, Rhizome	Extract	Anti-arthritic	[36]
Eleusine indica Gaertn	Wiregrass	Leaves	Oil	Anti-inflammatory, antioxidants, anti-arthritic	[39]
Imperata cylindrica (L.) Raeusch.	Thatch grass, Cogon grass	Rhizome, leaves, flowers	Decoction, powder	Diuretic, anti-inflammatory, antioxidant, immunomodulatory, neuroprotective	[40–42]
Lolium temulentum L.	Darnel	Seed	Powder	Anti-arthritic	[36]
Panicum italicum L.	Italian millet	Seed	Extract	Anti-arthritic	[35]
Setaria italic Beauv.	Foxtail millet	Grain	Parching	Anti-arthritic	[35]
Hordeum vulgare L.	Barley	Grain	Hydroalcoholic extract	Anti-chronic diseases, promote sleep, antidiabetes, enhance immunity, protective liver, beauty anti-acne, antioxidants, antidepressant, anticancer, improve gastrointestinal, anti-inflammation, hypolipidemic, preventive heart diseases, preventive constipation, improve cognition	[43, 44]
Triticum aestivum L. subsp. *spelta* (L.) Thell.	Wheat grass	Cotyledons, leaves	Extract, juice	Immunoprophylactic, anti-anemia, anti-leukemia, anti-inflammatory, diuretic, antioxidant, laxative, anti-aging anticarcinogenic; astringent, immunomodulatory, phosphorus, antibacterial, anti-venomous, sources of vitamins A and C, calcium, magnesium, potassium	[45, 46]

Table 8.
Forage grasses *and cereals specific medicinal properties.*

quadridentata seeds in Peru. *Grasses* stems contain sugar before the seeds mature, which gradually disappears. It is especially abundant in sorghum, corn, and optimal only in sugar cane (*Saccharum officinarum*). Fresh sugar cane contains 18% of his sugar weight. However, these odorless stems show aromatic properties in some grasses, such as *Cymbopogon schoenanthus* (sweet rush), *Anthoxanthum odoratum* (sweet rush), *Cymbopogon nardus* (*Ceylon citronella*), and *Cymbopogon citratus* (Citronella and Indian verbena).

Aromatic *Grasses* are known to be responsible for benzoic acid presence and an associated essential oil in herbivores urine. More specifically, *Saccharum fatuum* from Otahiti (French Polynesia) and *Bromus catharticus* from Peru are used to intoxicate fishes.

Grasses roots are sometimes used in medicine. Most are odorless, providing only little sugar and gum to the water. Main ones are: *Carex arenaria* (fake sarsaparilla), dogs: *Panicum repens* (Torpedo grass), *Imperata cylindrica* (Spear *grass*), *Cynodon dactylon* (Couch *grass*), and *Arundo donax* (cane root).

Some *Grass* roots are aromatic. This is the case with *Vetiveria odorata*, which contains a resin with a myrrh smell associated with volatile oil. *Vetiveria*' roots from Brazil served as a powerful sudorific.

Grass seeds provide a viscous drink in water; this seed decoction contains sugar, and gluten which dissolves with acetic and phosphoric acids. An herbal tea is made by mixing rice seed, oatmeal and barley. This drink contains grape sugar, dextrin, starch and gluten. The cane and quackgrass root are used to prepare soft drinks: quackgrass scales washed in cold water, contoured in a mortar and boiled for a quarter of an hour. Quackgrass (*Imperata cylindrica*) extracts are also obtained by roots leaching. *Grass starch* is used in many preparations to treat some humans and animals' diseases.

5.5 *Grasses and Grassland* ecological properties

Evidently, restoring native *Grasses* and *Grassland* is highly desirable. To that end, scientists might build models predicting human disturbance on global *Grassland* and assessing the climate-biosphere feedbacks as light grazing promoted soil C and N sequestration whereas moderate and heavy grazing significantly accelerated C and N losses. Indeed, light grazing also increase the above and belowground biomass, stimulate more fixed C allocated to roots and increase root exudates and biomass [47]. This enhances soil C accumulation as well N inputs into soils [48]. Meanwhile, light grazing also stimulate soil respiration by increasing temperature and moisture, enhancing ground cover, decreasing compaction, stimulating plant growth and microbial activities [49, 50]. However, both moderate and heavy grazing markedly decrease soil carbon pool and soil nitrogen pool as grazing decrease litter biomass, root C pool and microbial biomass and then lower C inputs to soils [51].

Fire remains a major disruption to evolution and management as well as determining *Grasses and Grassland* ecosystems. Fire is known for improving framework and resource environments, i.e., animal and livestock habitat. Prescribed fire is a tool for modern pasture management. Seasonal uses of fires, herbicide and nitrogen applications become promising, as desirable *grass* biomass increases while invasive plant biomass decreases. About three types of fires are mostly used in Tropics [52] for example, i.e. (a) *early fires* applied when the soil moisture degree is still sufficient to produce *grass* regrowth that is highly valued by livestock and covers their forage needs during the dry season. It cleans the straw left on the ground at the end of dry season. Its ignition date coincides with the end of the rainy season (mid-November to end of December, depending on the case). As results, it stimulates the growth of hemicryptophytes, improves primary biomass production of pastoral ecosystems and allows better land cover. (b) *Off-season fires* lit in the middle wet season and depends on: (i) the effort to conserve standing straw used as combustible for fire, (ii) the biomass ratio of "green matter/straw," which must be less than or equal to 1; where perennial *Grasses* abound, prescribed fires provide livestock with tender and palatable forage in a forage-deficit season. (c) *Late fires* applied when the degree of drying is maximum, very violent, compromising the regeneration of forest recruits and chamephytes species often despised by animals. It reduces woody plants density, promotes *Grass* growth, and accelerates bare beaches reducing by the way the pastures' carrying capacity.

But, using fire for sustainable *Grasses and Grassland* require optimal ecological conditions as well as specific well trained staffs.

Some species are important in regards to their roots which were highly used for human and animal medicine. These species (*Andropogon* spp., *Cenchrus* spp., *Cynodon dactylon* and *Imperata cylindrica*) appeared to be highly threatened and thus, their culture should be encouraged in order to make them more available for all needs.

It appears that all the listed *Grasses* are heavily attacked by a multitude of diseases and parasites (**Table 9**). Although efforts have been made to enable each of these plant species to provide the expected yields (productivity, food quality …), from another point of view, it is important to highlight their overall major ecological

Grasses	Model species	Number diseases	Main foliar diseases	Economic parasites	References
Wheat-grasses	*Agropyron* spp. *Hordeum* spp. *Triticum* spp.	70	Rusts, smuts, blotches, spots, scald, scolecotrichum stripe, powdery mildew	*Erisyphe graminis*	[53]
Needle-grasses	*Stipa* spp.	>50	Spots (septoria, selenophoma, stem), stagonospora .blotch, stripes	*Scolecotrichum graminis*	[53]
Blue-grasses	*Poa* spp.	50	Rusts, stripe, spots, powdery mildew, anthracnose, smut, fairy ring, melting-out, silvertop, slime molds	*Laetisaria spp, Entyloma spps, Clitocybe, Lycoperdon, Psalliota, Typhula, mites, Drechslera, Fusarium, Erysiphe graminis*	[53]
Smutgrasses dropseeds	*Sporobolus* spp.	36	Rusts, spots, powdery mildew, mold, Smut	*Helminihosporium ravenelii*	[54]
Grama grasses	*Bouteloua* spp. *Aristida* spp.	30	Rust, spots, choke, scald, black ring	*Balansia strangulans, Balansia hemicrypta*	[55]
Canary grasses	*Phalaris* spp.	30	Tawny spot on leaves	*Pseudomonas coronajaciens*	[53]
Brome grasses (Mountain, meadow, smooth)	*Bromus* spp.	38	Spots, anthracnose, powdery mildew, mold, scalds, rot, speckle, blotch, stripes, rusts, Smut, Silvertop, Blight, Node	*Pseudomonas, Ustilago bullata, Fusarium, Pyrenophora sp, Selenophoma bromigena, Xanthomonas translucens, Alternaria spp., Claviceps spp, Erysiphe graminis*	[56]

Grasses	**Model species**	**Number diseases**	**Main foliar diseases**	**Economic parasites**	**References**
Foxtail Millet	*Setaria itálica*	22	Spots (Cercospora, helminthosporium, bacterial, gray), downy mildew	*Piricularia grísea, Sclerospora graminicola*	[53]
Meadow Foxtail	*Alopecurus pratensis*	19	Stripe, water-soaked, blotches, scalds, streak, rust	*Rhynchosporium spp, Drechslera sp., Sclerotinia borealis; Puccinia graminis*	[53, 56]
Buffalo grass	*Buchloë dactyloides*	11	Spots, false smut	*Cercospota seminalis*	[53]
Beardgrass/ Bluestem	*Andropogon* spp.	—	Rusts, choke, anthracnose, cattail, spots	*Phyllachora luteomaculata*	[57]
Wild-rye grasses Junegrass	*Elymus* spp. *Koeleria cristata*	14	Rusts, smuts, powdery mildew, spots, blight, stripe, speckle	*Epichloe typhina*	[53]
Bentgrass	*Agrostis* spp.	13	Patchs, molds, spots, leaf blotch, Pythium Blight, red thread	*Rhizoctonia solani, Typhula spp, Coprinus spp, Drechslera, Sclerotinia, Leptosphaeria korrae, Fusarium, Pythium, Laetisaria spp*	[58]
Fescue, Red	*Festuca rubra*	14	Anthracnose, molds, red thread, silver top, spots, scald, smuts, Brown stripe, blotch, powdery mildew, rust	*Colletotrichum graminicola, Typhula spp, Coprinus spp, Fusarium spp, Clitocybe spp Marasmius spp, Lycoperdon spp, Psalliota*	[59]
Timothy grass	*Phleum* spp.	06	Spot, blight, rot, mold, ergot, snow scald	*Heterosporium phleum, Drechslerai, Pythium, Claviceps purpurea, Sclerotinia boreali*	[60]

Table 9.
Grasses and Grassland *pressures and extents of diseases.*

interest. Indeed, these species also serve as shelters and refuges or habitats as well as food resources for many of these parasites, i.e., insects, bacteria, fungi, etc. They therefore participate in the development of ecological niches in a context of habitat fragmentation, an overall major role. Therefore, how can we perpetuate the usefulness of *Grasses* in the midst of various plagues and attacks from grasshoppers, striga, cantharids and anthracnose?

Conservation and enhancement for the majority of introduced forage species are aimed at scientific, cultural, and touristic purposes. Herbarium and *ex situ* conservation, i.e., laboratories, are still on a very small scale in Central and West Africa.

Some tropical native *Grasses* deserve special attention such as:

- Fire-sensitive species that, lose fruit-bearing performance. These are therophytes *Grasses*, mainly *Hyparrhenia involucrata* and *Pennisetum polystachion* [61].

- Forage sensitive to trampling, i.e., *Aristida kerstingii*, *Urochloa indica*.
- Rare genera: *Elymandra androphyla, Loxodera ledermannii, Hyperthelia dissoluta*, which require further attention
- Grazing-sensitive including *Hyparrhenia smithiana* and *Pennisetum unisetum* [3, 4, 9], *Setaria sphacelata, Anadelphia afzeliana, Brachiaria falcifera,* and *Loxodera ledermannii,* less silicified under grazing pressure making them more vulnerable [12].
- Site indicator, i.e., *Brachiaria brachylopha* for dry sites and *B. falcifera* for subhumid sites [62].

Grasses and Grassland biodiversity conservation requires (i) a comprehensive census to define the biodiversity conservation strategy; (ii) programming and planning to regulate grazing pressure; (iii) updating revision of herbaria; (iv) *in situ* conservation, and (v) wild *Grasses* domestication.

However, a specialist look is requiring for following concerns: (i) Are livestock grazing and burning compatible? (ii) How to reduce invasive species impact on *Grasses and Grassland* facing biodiversity erosion? (iii) What alternatives to issues in temperate environments as well as facing pests' damage extent? (iv) What would be *Grasses and Grassland'* contribution to agricultural and environmental services in a variety of systems that value permanent *Grassland* in the forage system?

5.6 *Grasses and Grassland* cultural properties

Grasses are important both in everyday life and during mourning ceremonies, links, knot of all kinds (portage, forbidden, calendars, stubble or signs of fallow land ...) [63]. They are surrounded by several spells, superstitions, myths, legends as well as popular beliefs. Among 15 of these properties reported for example on the wheat are the following:

- *Wheat is a tool in religious rites* of ancient Egyptians, Assyrians, Chaldeans, Romans, Greeks, as well as in India. Its bread has become the central mystery of Christianity. As a symbol of food, fertility and the annual rebirth of life, wheat is the offering of the Harvest Day in Luna sad, Ireland (August 1 in the Northern Hemisphere, February 1 in the Southern Hemisphere).
- *Wheat is a symbol of fertility* often used newlyweds' home decoration [64], offering happy inspiration, gratitude, prudence united to goodness, legitimate acquisition [65].
- *Lucky sheaves*: for hares, partridges and farmers because of the Wheat genius which embodies a last sheaf shaped like a wolf.
- *Uncompromising guardians*: Other wheat geniuses such as *Polievik* in Russia or *Polevik* in Poland hunt pests and weeds and promote harvests. But he will strangle those who would take a nap instead of plowing their land! *Poludnica* ensures breaks observance. *Polednice* in Czechoslovakia and Slovakia prevents damage before harvest while *Polednicek* prevents fields from being ransacked.

6. *Grasses and Grassland* syntheses

Grasses hold a lot of syntheses as about 27 have been partly listed from the main cereals (sorghum, oat, barley, wheat, maize, Spelt). These chemicals and biologicals act as anticancer [37, 66], increasing milk production [67], antifungal, antibacterial [37], anthelmintic, anti-amoebic, cyclooxygenase (COX) I and II inhibitory activity [68], dysmenorrheal and uterine relaxing activities.

Among the *Sorghum* synthesis (**Table 10**), very little is still known on the Dhurrin genetic control. However, Dhurrin content was recently found in consistent association with biosynthetic genes in N-fertilized environments, while with catabolic loci in the controls [78]. Several phenolic compounds also accumulated radioactivity.

Engineering strategies targeting plant biomass lignin develop sustainable bio economy. Tricin native monocots lignin polymer initiate the lignin chains polymerization. Its bio-synthesis requires two methylation reactions involving the pathway intermediate selgin. O-methyltransferase is producing S lignin units as in the lignin-linked tricin synthesis [79].

The 1,3-propanediol is bio-produced from white sorghum starch and glycerol inoculated by a mixture culture of *Escherichia coli* and *Klebsiella* species [80].

Meeting the human nutrition balance needs require improving protein and amino acids relative contents in rice grain.

The use of genetic engineering strategy can improve essential amino acids contents, and nutritional quality of rice grain. But, how to regulate lipid metabolism pathway in rice grains remains questionable.

Species	Bio-synthesis	Characteristics	References
Sorghum	Dhurrin	• Cyanogenic glucoside in early grain development • Low-juveniles vs. high-older plants contents • Highest in *Caudatum* and lowest in *Guinea* • Shift from leaves to stem	[13]
Rice	Amino acid in grains	• Lysine, first limiting amino acid in cereals • Lysine, methionine, threonine and isoleucine	[69]
	Aromatic amino-acids	• Increasing tryptophan and phenylalanine rates • Cysteine, serine, methionine change into isoleucine	[70]
	Vitamins in grains	• Foliar β-carotenoid synthesize humans vitamin A • Transgenic produce β-carotenoids in endosperm. • Transgenic folic acid as 1.5 times of original • Thiamine/vitamin B1 low in plastids causing beriberi • Vitamin E as tocopherol, trienol family • Vitamin E as anti-oxidative damage protection • Vitamin C as antioxidant, antiatherosclerosis, anti-cancer	[71, 72]
Oat	Avenancins as saponins	• Protective properties, i.e. anti-inflammatory, antifungals, anti-bacterial, anti-parasitism, anti-cancers and anti-viral	[73]
Wheat	Storage protein	• Prolamin gene stimulation at storage onset	[74]
	Gluten	• Gluten proteins heredity by starch gel electrophoresis • No gluten effect on Canthatch variety D genome	[75]
	Antioxidants	• Antioxidant by reducing glycoside, polyphenolics	[76]
Barley	Starch and amino acid in grains	• Less starch, dry weight in transgenic endosperms	[77]

Table 10.
Some cereals synthesis.

Little is known about Vitamin C biosynthesis in monocotyledonous plants. Therefore, solving low vitamin content in rice grains, require increasing vitamin of rice seeds and improving nutritional quality by making rice, full use of its genes. Achieving this might consider the introduction of exogenous genes, metabolic engineering, genetic engineering and other modern technical methods.

7. Grasses and Grassland applications

Ecosystem services of *Grasses and Grassland* are expressed through grazing areas, watershed water, biodiversity reserves, tourist sites, recreation areas, religious sites, wild food sources, and natural medicine sources, mainly through the sequestration and storage of C.

Several species of large *Grasses* are also very important for the manufacture of mats and for the roof. Normally, they are protected against fire and grazing, but rarely cultivated. For the manufacture of mats along the Logone, *Hyparrhenia rufa* is planted. *Vetiveria nigritana* is used for the fields' demarcation on floodplains, and anti-erosion. Vetiver's thatch is traditionally used to cover the roofs of straw huts, and to make basketry or carpets. *Bambusa vulgaris* is planted for timber in southern countries while *Oxytenanthera abyssinica* is often more notable in the north.

Miscanthus from Asia as well as sweet sorghum (*Sorghum bicolor*) and switch-grass (*Panicum virgatum*) from USA [81] are material for biofuel and buildings. Ciments Calcia and Alkern as Industrial company substitute's traditional aggregates with crushed to reach 60% *Miscanthus* in the concrete; a prototype block of $20 \times 50 \times 20$ cm weighing 17 kg is three times more insulating than conventional concrete [82]. Insulating panels and biomaterial in term of acoustic comfort with noise attenuation, it is 4 h fire resistance. First experimental use of concrete in early 2018 deployed on 1700 m^2 of facade of 46 social housing units in Chanteloup-en-Brie (France) requires 50 tons of *Miscanthus*.

These are biosourced materials, grown without pesticides and irrigation, adapts to polluted, degraded or abandoned land, out of competing food agriculture, offering additional resources and economic opportunities for farmers. Production is spread over 15–20 years without reseeding or fertilizing. In addition, *Miscanthus* is sterile, rhizome and non-invasive, yielding 10 tons/ha per year. It reduces the building's carbon footprint by saving on the transport of aggregates over long distances.

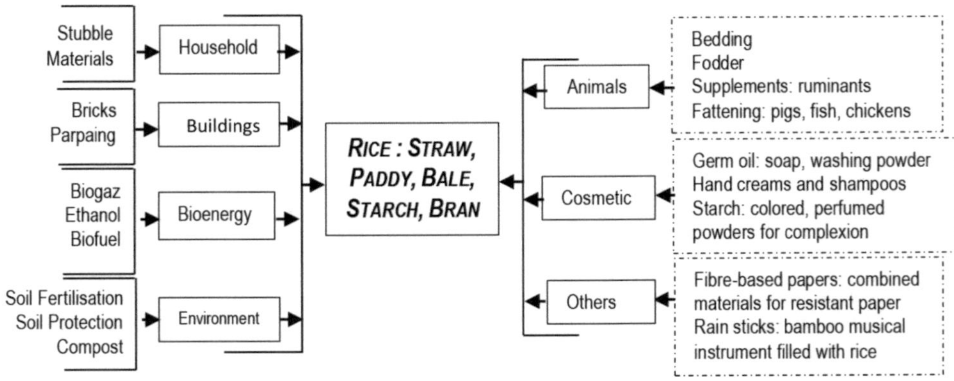

Figure 3.
Applications from the rice.

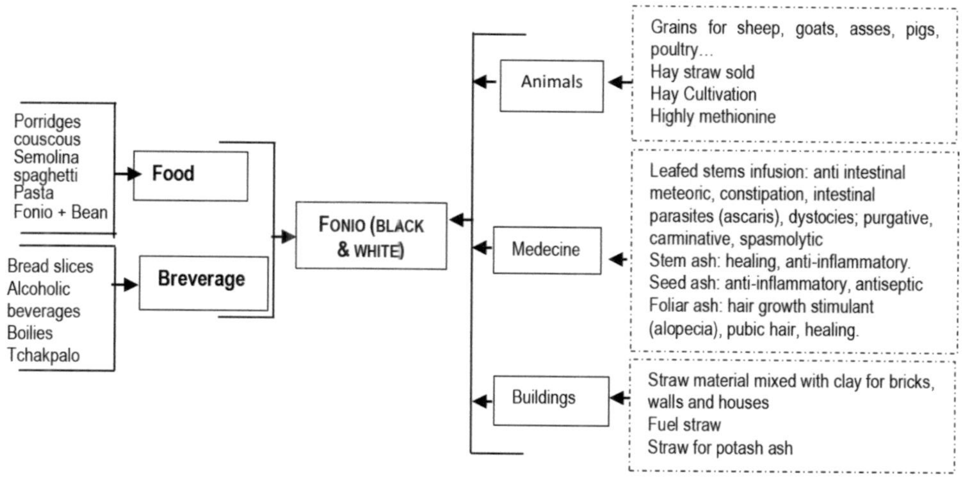

Figure 4.
Applications from the Fonio.

Arundo donax provides all-lignocellulosic fiberboards without synthetic binders, raw good material for fiberboard production and its pulp is rich in cellulose and moderate in lignin.

Figures 3 and **4** highlight some applications from Rice and Fonio, respectively.

8. Concluding remarks

This chapter highlights some milestones in *Grasses and Grassland* research: (1) 18 types of *Grassland* identified; (2) only 250 tropical *grass* species among 9700, i.e., 2.58% studied before 1990 and 100 over the last 30 years; (3) *Grasses* have been field characterized throughout 78 parameters; (4) biodiversity results in 43 cereals and 50 tropical *Grasslands*; (5) mostly, *Grassland* of *Andropogon gayanus* and *Hyparrhenia involucrata* (74% PV, 6.28 tons DM/ha; 0.37 ha/TLU) from Guineo-Sudanese opposed *Setaria longiseta* and *Sporobolus pyramidalis* (21% PV; 2.9 tons DM/ha; 5.4 ha/TLU) from Sudanian zone; (6) cultivated *grasses* belong to *Panicum* spp., *Brachiaria* spp. and *Pennisetum* spp.; (7) cereals contain gluten from 5% in rice to 69% in wheat; (8) properties result in 7 anti-erosive, 10 ornamental, 17 medicinal, 15 cultural and 11 aromatic with 95 oil volatile compounds; (8) Twenty-seven bio-syntheses recorded with Dhurrin genetic control, mechanism of lipid metabolism pathway and vitamin C biosynthesis remaining concerns; (9) applications are bioenergetic, cosmetic, industrial and environmental with sequestering carbon and nitrogen into soil.

Otherwise, most of cereals' syntheses are found in fighting century diseases including cancers, high blood pressure, etc. that devastate human resources and thus negatively impact Nations economies.

So many virtues for more or less demanding resources like *Grasses and Grassland*! The whole world can certainly make a big profit from its. In a context of increasing difficulties in adequately feeding a growing world population threatened by major plagues, interest in these natural resources must certainly be attracted. Given the rapid precariousness of the food and health situation in some parts of the world, would it be too much to consider *Grasses and Grassland* as a hope for sustainable well-being? Whether ecological intensification is a pastoral contribution to agricultural and environmental services in a variety of systems that value permanent *Grasslands* in the forage system, what would life on earth really be without these *Grasses and Grassland*? Facing array of properties, synthesis and applications which force the hope, politicians would science-based manage for ensuring secured future for humanity, even in the increasingly alarming global warming; because *Grasses and Grassland* could be solutions to eventualities. In this debate, however, government officials, policy makers, professionals, and the general public would ensure that proactive and sustained production, processing and development, as well as commercialization of *Grasses and Grassland* are for the sustainable well-being of the respective communities; Decisions needed for sustainably managing these properties, synthesis, and applications so that serious threats can be mitigated. This includes students, teachers, and operators, who are tracking accurate and updated inventories of *Grasses and Grassland*' knowledge.

Acknowledgements

This research was co-funded by the Laboratory of Applied Ecology—LEA (University of Abomey-Calavi, Benin—UAC), the Ecological and Organic Agriculture Network and the Association Béninoise pour le Pastoralisme (ABEPA). I am acknowledging my Master Professor Brice Sinsin, Former UAC Rector and Head of LEA for mentoring, DiasporaEngager (Georgia, USA) and it CEO, Roland Holou (PhD Agronomy) and Ozias Hounkpatin (PhD Agronomy) from Uppsala University (Sweden) for friendship and paper review, Bishop Barthelemy Tiando Bona and Pastor Arnaud Assogba (BSc Sociology) for prayers. Irma Cale Pascaline Kpakpo-Kindomihou (BSc Accounting) provides good working atmosphere while Iva Simcic, Ivana Spajic and Edi Lipovic' collaborative efforts, timely boost in publishing this chapter.

Author details

Valentin Missiakô Kindomihou
Laboratory of Applied Ecology, Faculty of Agronomic Sciences, Republic of Benin, West Africa

*Address all correspondence to: kindomihou@gmail.com; valentin.kindomihou@fsa.uac.bj

References

[1] Pottier E, Michaud A, Farrié JP, Plantureux S, Baumont R. Les prairies permanentes françaises au cœur d'enjeux agricoles et environnementaux. Innovations Agronomiques. 2012;**25**: 85-97

[2] Meynard JM. Introduction générale—Produire autrement: réinventer les systèmes de culture, Systèmes de culture innovants et durables: Quelles méthodes pour les mettre au point et les évaluer? Dijon: Educ agri éd.; 2008. pp. 11-27

[3] Sinsin B. Phytosociologie, écologie, valeur pastorale, production et capacité de charge des pâturages naturels du périmètre Nikki-Kalalé au Nord-Bénin [thesis]. Belgium: Université Libre de Bruxelles; 1993

[4] Sinsin B. Dans le Parc National de la Pendjari, la prise en compte des plantes herbacées est nécessaire pour définir valablement les associations végétales [thesis index]. Belgium: Université Libre de Bruxelles; 1993

[5] Balay C, Cournut S, Michelin Y, Capitaine M, Boisdon I. Services écosystémiques rendus par les prairies dans une commune de moyenne montagne d'Auvergne: évaluation et déterminants. Fourrages. 2015;**221**: 15-24

[6] Skerman PJ, Riveros F, editors. Tropical grasses. In: FAO Plant Production and Protection Series. No. 23. 1990. 832p

[7] Kindomihou MV, Sinsin B, Holou RYA, Ambouta J-MK, Gruber W, Adjolohoun S, et al. The effect of seasonal variations, covariations with minerals and forage value on Itchgrass' foliar silicification from Sudanian Benin. SILICON. 2016;**8**(4): 487-496. DOI: 10.1007/s12633-015-9355-y

[8] Kindomihou VM, Oumorou M, Mensah GA, Sinsin BA. Morphological traits and germination of *Loxodera ledermannii* (Pilger) W.D. Clayton caryopses in southern-Benin. Bulletin de Recherche Agronomique du Bénin. 2009;**65**:37-43

[9] Sinsin B. Individualisation et hiérarchisation des phytocénoses soudaniennes du Nord-Bénin. Belgian Journal of Botany. 1994;**127**(1):87-103

[10] Klein HD, Rippstein G, Huguenin J, Toutain B, Guerin H, Louppe D. Les cultures fourragères. Quae, CTA, Gembloux: Presses agronomiques de Gembloux; 2014. 264 p

[11] Kindomihou V, Adandédjan C, Sinsin B. Performances agronomiques et zootechniques d'associations d'espèces fourragères tropicales au nord-Bénin (Zone soudanienne). In: Godet G, Grimaux P, Guerin H, editors. Actes de l'Atelier Régional sur "Cultures Fourragères et Développement Durable en Zone Humide". Côte d'Ivoire: CIRDES/IDESSA Korhogo; 1998. pp. 80-85

[12] Kindomihou MV. Tropical *grasses silicification:* Genetic interspecific variation, influence of growth conditions and relationship with the foliar structure [thesis]. Belgium: Free University of Brussels; 2005

[13] Kindomihou MV. Sorghum: Properties, Synthesis and Applications. Serie Agriculture, Issues and Policies. New York, USA: Nova Science Publishers Inc.; 2019 319p

[14] Dagbenonbakin GD, Kindomihou V, Agbangba EC, Sokpon N, Sinsin B. Diagnosis and recommendation integrated system (DRIS) model establishment for diagnosing *Sorghum* (*Sorghum bicolor*) nutrient status in Benin (West Africa).

Scientific Research and Essays. 2013; **8**(25):1562-1569. DOI: 10.5897/SRE11

[15] Kindomihou V. Diversité des plantes fourragères et des pâturages naturels au Bénin. In: Rapport de Mission. Cotonou, Bénin: Projet Stratégie Nationale Biodiversité/MEHU; 2000 42 p

[16] Houinato MBR. Phytosociologie, écologie, production et capacité de charge des formations végétales pâturées dans la région des Monts Kouffé (Bénin) [thesis]. Belgium: Université Libre de Bruxelles; 2001

[17] Natta AK. Ecological assessment of riparian forests in Bénin: Phytodiversity, phytosociology and spatial distribution of tree species [thesis]. Holland: Wageningen University; 2003

[18] Aboh A. Phytosociologie, écologie, potentialités et aménagement des pâturages naturels envahis par *Chromolaena odorata* et *Hyptis suaveolens* en Zone Soudano-guinéenne (Bénin) [thesis]. Benin: Université d'Abomey-Calavi; 2008

[19] Djenontin J. Dynamique des stratégies et des pratiques d'utilisation des parcours naturels pour l'alimentation des troupeaux bovins au Nord-Est du Bénin [thesis]. Benin: Université d'Abomey-Calavi; 2010

[20] Sinsin B, Owolabi B. Monographie nationale de la biodiversité. In: Rapport de Mission. Cotonou, Bénin: Projet Stratégie Nationale Biodiversité/MEHU; 2002 42p

[21] Seignobos C, Donfack P. Des plantes indicatrices dans un agrosystème incluant la jachère: les exemples des Peuls et des Giziga du Nord-Cameroun. Journal d'Agriculture Traditionnelle et de Botanique Appliquée. 1996;**18**(1): 231-250

[22] Kindomihou V, Téka O, Houinato M, Oumorou M, Mama A, Ogoubiyi V, et al. Climate change and in situ conservation of *Loxodera ledermannii* (Pilger) natural pastures in Sudanian Benin. In: Halford et al., editors. "Effects of Climate Change on Plants: Implications for Agriculture". Aspects of Applied Biology. Vol. 88. 2008

[23] Shah G, Shri R, Panchal V, Sharma N, Singh B, Mann AS. Scientific basis for the therapeutic use of *Cymbopogon citratus*, stapf (lemon *grass*). Journal of Advanced Pharmaceutical Technology & Research. 2011;**2**(1):3-8. DOI: 10.4103/2231-4040.79796

[24] Boukhatem MN, Ferhat MA, Kameli A, Saidi F, Kebir HT. Lemon *grass* (*Cymbopogon citratus*) essential oil as potent anti-inflammatory and antifungal drugs. Libyan Journal of Medicine. 2014;**9**. DOI: 10.3402/ljm.v9.25431

[25] Prates HT, Leite RC, Craveiro AA, Alaíde BO. Identification of some chemical components of the essential oil from molasses *grass* (*Melinis minutiflora* Beauv.) and their activity against cattle-tick (*Boophilus microplus*). Journal of the Brazilian Chemical Society. 1998;**9**(2): 193-197. DOI: 10.1590/S0103-50531998000200013

[26] Chahal KK, Urvashi B, Kauchal S, Sandhu AK. Chemical composition and biological properties of *Chrysopogon zizanioides* (L.) Roberty Syn. (L.) Nash. A review. Indian Journal of Natural Products and Resources. 2015;**6**(4): 251-260

[27] Choi N, Lee JS, Kwak J. Production of biodiesel from acid oil via a two-step enzymatic trans-esterification. Journal of Oleo Science. 2016;**65**:913-921

[28] Ramos DM, Diniz P, Ooi MKJ, Borghetti F, Valls JFM. Avoiding the dry

season: Dispersal time and syndrome mediate seed dormancy in *Grasses* in Neotropical savanna and wet *grasslands*. Journal of Vegetation Science. 2017;**28**: 798-807. DOI: 10.1111/jvs.12531

[29] Desseaux A. Aromathérapie en cancérologie: rationnel, intérêt et limites. Sciences Pharmaceutiques. 2018. 391 p. dumas-01928000

[30] Peterson PM, Planchuelo AM. *Bromus catharticus* in South America (Poaceae: Bromeae). Novon. 1998;**8**(1): 53-60. DOI: 10.2307/3391893

[31] Abbott LA, Pistorale SM, Filippini OS. Path coefficient analysis for seed yield in *Bromus catharticus*. Ciencia e Investigación Agraria. 2007; **34**(2):107-114

[32] Cornu A, Carnat AP, Martin B, Coulon JB, Lamaison JL, Berdague JL. Solid phase microextraction of volatile components from natural *grassland* plants. Journal of Agricultural and Food Chemistry. 2001; **49**:203-209

[33] Tornambe G, Cornu A, Pradel P, Kondjoyan N, Carnat AP, Petit M, et al. Changes in terpene content in milk from pasture-fed cows. Journal of Dairy Science. 2006;**89**:2634-2648

[34] Tornambe G, Cornu A, Verdier-Metz I, Pradel P, Kondjoyan N, Figueredo G, et al. Addition of pasture plant essential oil in milk: Influence on chemical and sensory properties of milk and cheese. Journal of Dairy Science. 2008;**91**(1):58-69. DOI: 10.3168/jds.2007-0154

[35] Nadkarni KM, Nadkarni AK. Indian Material Medica. 3rd ed. Mumbai: Popular Prakashan; 2005

[36] Bentley R, Trimen H. Medicinal Plants. Delhi, India: Asiatic Publishing House; 2002.Reprint Indian Edition

[37] Alothman EA, Awaad AS, Al-Qurayn NA, Al-Kanhal HF, El-Meligy RM, Zain YM, et al. Anticancer effect of *Cenchrus ciliaris* L. Saudi Pharmaceutical Journal. 2018;**26**:952-955. DOI: 10.1016/j.jsps.2018.05.003

[38] Sharma RK, Singh B, Sahoo A. Exploring feeding value of oak (*Quercus incana*) leaves: Nutrient intake and utilization in calves. Livestock Science. 2008;**118**(1/2):157-165

[39] Sagnia B, Fedeli D, Casetti R, Montesano C, Falcioni G, Colizzi V. Antioxidant and anti-inflammatory activities of extracts from *Cassia alata, Eleusine indica, Eremomastax speciosa, Carica papaya* and *Polyscias fulva* medicinal plants collected in Cameroon. PLoS One. 2014;**9**(8):e103999

[40] Yue X-R, Hou Z-X, Liu P, Wang S-S. Anti-inflammatory effect of imperata cylindrical. Chinese Journal of Clinical Rehabilitation. 2006;**10**(43): 85-87

[41] Kwok AH, Wang Y, Ho WS. Cytotoxic and pro-oxidative effects of *Imperata cylindrica* aerial part ethyl acetate extract in colorectal cancer in vitro. Phytomedicine. 2016;**23**(5): 558-565. DOI: 10.1016/j.phymed.2016.02.015

[42] Anggraeni N, Syamsunarno MRAA, Widyastuti R, Puspitasari IM, Praptama S. Potential dual effect anti-inflammatory and anti-platelet of cogon grass ethanol extract on diabetic mice a preliminary study. IOP Conference Series: Journal of Physics. Conference Series. 2019: 1246. DOI: 10.1088/1742-6596/1246/1/012006

[43] Kubatka P, Kello M, Kajo K, Kruzliak P, Výbohová D, Šmejkal K, et al. Young barley indicates antitumor effects in experimental breast cancer in vivo and in vitro. Nutrition and Cancer. 2016;**68**(4):611-621

[44] Zeng Y, Pu X, Yang J, Du J, Yang X, Li X, et al. Preventive and therapeutic role of functional ingredients of barley *grass* for chronic diseases in human beings. Oxidative Medicine and Cellular Longevity. 2018:1-15. ID 3232080. DOI: 10.1155/2018/3232080

[45] Khan N, Ganeshpurkar A, Dubey N, Bansal D. Immunoprophylactic potential of wheat *grass* extract on benzene-induced leukemia: An in vivo study on murine model. Indian Journal of Pharmacology. 2015;**47**(4):394-397. DOI: 10.4103/0253-7613.161261

[46] Das P, Mandal S, Gangopadhyay S, Das K, Mitra AG, Dasgupta S, et al. Antioxidative and anticarcinogenic activities of methylpheophorbidea, isolated from wheat *grass* (*Triticum aestivum* Linn.). Natural Product Research. 2016;**30**(4):474-477. DOI: 10.1080/14786419.2015.1022775

[47] Liu N, Kan HM, Yang GW, Zhang YJ. Changes in plant, soil, and microbes in a typical steppe from simulated grazing: Explaining potential change in soil C. Ecological Monographs. 2015;**85**:269-286

[48] Derner J, Briske D, Boutton T. Does grazing mediate soil carbon and nitrogen accumulation beneath C_4, perennial *grasses* along an environmental gradient? Plant and Soil. 1997;**191**: 147-156

[49] Gong JR, Wang YH, Liu M, Huang YM, Yan X, Zhang ZY, et al. Effects of land use on soil respiration in the temperate steppe of Inner Mongolia, China. Soil and Tillage Research. 2014; **144**:20-31

[50] Zhang T, Zhang YJ, Xu MJ, Zhu JT, Wimberly MC, Yu GR, et al. Light-intensity grazing improves alpine meadow productivity and adaption to climate change on the Tibetan plateau. Scientific Reports. 2015;**5**:15949

[51] Knops JMH, Bradley KL, Wedin DA. Mechanisms of plant species impacts on ecosystem nitrogen cycling. Ecology Letters. 2002;**5**:454-466

[52] Teka O, Houessou L, Kindomihou V, Sinsin B. Use of vegetation fires as tool in pastoral land management. In: Sinsin B, Kampmann D, editors. Biodiversity Atlas of West Africa. Vol. 1. Benin, Cotonou & Frankfurt/Main; 2011. pp. 196-203. ISBN 978-3-9813933-0-9 (hardcover); ISBN 978-3-9813933-3-0 (paperback)

[53] Hardison JR. Role of fire for disease control in *grass* seed production. Plant Disease. 1980;**64**:641-645

[54] Fushtey SG, Taylor DK, Fairey DAPHNE. The effect of wear stress on survival of turfgrass in pure stands and in mixtures. Canadian Journal of Plant Science. 1983;**63**(1): 317-322. DOI: 10.4141/cjps83-033

[55] Gray SM, Banerjee N, Thomas JS, Chapin JW, Smith DM. Barley yellow dwarf luteoviruses and their predominant aphid vectors in winter wheat grown in South Carolina. Plant Disease. 1998;**82**:1328-1333

[56] Turnbull GD, Gossen BD. Head smut of *grasses* on the Canadian prairies. II. Host range and variability. Canadian Journal of Plant Pathology. 1996;**18**(3): 255-260. DOI: 10.1080/07060669609500621

[57] Soroka JJ, Gossen BD. Phytophagus arthropods and silvertop levels associated with post-harvest residue treatments in three turfgrass species grown for seed. Canadian Journal of Plant Science. 2005, 2001;**85**:213-224

[58] Smith JD, Gossen BD, Hsiang T. First report of dollar spot, caused by *Sclerotinia homoeocarpa*, on *Poa pratensis* in Saskatchewan, Canada. Plant

Diseases. 2001;**85**(7):803. DOI: 10.1094/PDIS.2001.85.7.803A

[59] Hsiang T, Mahuku GS. Genetic variation within and between southern Ontario populations of *Sclerotinia homoeocarpa*. Plant Pathology. 1999;**48**: 83-94. DOI: 10.1046/j.1365-3059.1999.00306.x

[60] Smith D. Influence of cool and warm temperatures and temperature reversal at inflorescence emergence on yield and chemical composition of timothy and bromegrass at anthesis. In: Proceedings of the 11th International Grassland Congress. 1970. pp. 510-514

[61] Sinsin B, Saidou A. Impact des feux contrôlés sur la productivité des pâturages naturels des savanes soudano-guinéennes du Ranch de l'Okpara au Bénin. Annales des Sciences Agronomiques du Bénin. 1998;**1**:11-30

[62] Sinsin B, Essou JP, Saidou A, Houinato M, Kindomihou V, Bako I, et al. Gestion des pâturages naturels de la ferme d'élevage de Bétécoucou par le feu. In: Rapport principal. Cotonou: MDR/DE/PDPA/LEA; 1996. 52 p

[63] La Fontinelle J. Mauvaises herbes et graminées à Houailou (Nouvelle-Calédonie). Journal de la Société des Océanistes. 2002;**114–115**:39-43. DOI: 10.4000/jso.1370

[64] Carr-Gomm P, Carr-Gomm S. L'Oracle druidique des plantes. Edition Vega.; 2006. p. 144

[65] Romey G. Dictionnaire de la Symbolique, le vocabulaire fondamental des rêves, Tome 1: couleurs, minéraux, métaux, végétaux, animaux. Canada: Edition Albin Michel; 1995. p. 576

[66] Alves VG, Souza AG, Chiavelli LU, Ruiz AL, Carvalho JE, Pomini AM, et al. Phenolic compounds and anticancer activity of commercial sugarcane cultivated in Brazil. Anais da Academia Brasileira de Ciências. 2016;**88**(3): 1201-1209. DOI: 10.1590/0001-3765201620150349

[67] Marshall J. Infant feeding: Anatomy and physiology. The Practising Midwife. 2012;**15**:38-41

[68] Light ME, McGaw LJ, Sparg SG, Jäger AK, Staden J. Screening of *Cenchrus ciliaris* L. for biological activity. South African Journal of Botany. 2002;**68**:411-413

[69] Osuji GO, Duffus E, Johnson P. Enhancement of the essential amino acid composition of food crop proteins through biotechnology. American Journal of Plant Sciences. 2015;**6**: 3091-3108

[70] Birla DS, Malik K, Sainger M. Progress and challenges in improving the nutritional quality of rice (*Oryza sativa* l.). Critical Reviews in Food Science and Nutrition. 2017;**57**: 2455-2481

[71] Headey D, Hoddinott J, Ali D. The other Asian enigma, explaining the rapid reduction of under nutrition in Bangladesh. World Development. 2015; **66**:749-761

[72] Naidu KA. Vitamin C in human health and disease is still a mystery? An overview. Nutrition Journal. 2003;**2**:7

[73] Bach TJ, Rohmer M. Isoprenoid Synthesis in Plants and Microorganisms: New Concepts and Experimental Approaches. 2013. DOI 10.1007/978-1-4614-4063-5_28

[74] Weichert N, Saalbach I, Weichert H, Kohl S, Erban A, Kopka J, et al. Increasing sucrose uptake capacity of wheat grains stimulates storage protein synthesis. Plant Physiology.

2010;**152**:698-710. DOI: 10.1104/pp.109.150854

[75] Boyd WJR, Lee JW. The control of wheat gluten synthesis at the genome and chromosome levels. Experientia. 1967;**23**:332-333. DOI: 10.1007/BF02144496

[76] Calzuola I, Marsili V, Gianfranceschi GL. Synthesis of antioxidants in wheat sprouts. Journal of Agricultural and Food Chemistry. 2004;**52**(16):5201-5206. DOI: 10.1021/jf0307752

[77] Radchuk V, Riewe D, Peukert M, Matros A, Strickert M, Radchuk R, et al. Down-regulation of the sucrose transporters HvSUT1 and HvSUT2 affects sucrose homeostasis along its delivery path in barley grains. Journal of Experimental Botany. 2017;**68**(16):4595-4612. DOI: 10.1093/jxb/erx266

[78] Nielsen LJ, Stuart P, Pičmanová M, Rasmussen S, Olsen CE, Harholt J, et al. Dhurrin metabolism in the developing grain of *Sorghum bicolor* (L.) Moench investigated by metabolite profiling and novel clustering analyses of time-resolved transcriptomic data. BMC Genomics. 2016;**17**:1021. DOI: 10.1186/s12864-016-3360-4

[79] Eudes A, Dutta T, Deng K, Jacquet N, Sinha A, Benites V, et al. SbCOMT (Bmr12) is involved in the biosynthesis of tricin-lignin in *Sorghum*. PLoS One. 2017;**12**(6):e0178160. DOI: 10.1371/journal.pone.0178160

[80] Obele CM, Ogbobe O, Okonkwo IF. Synthesis of 1,Z-propanediol from *Sorghum*. Pakistan Journal of Nutrition. 2010;**9**(11):1058

[81] Holou R, Kindomihou VM. The biofuel crops in global warming challenge: Carbon capture by corn, sweet sorghum and switchgrass biomass grown for biofuel production in the USA. In: Jacob-Lopes E, editor. Frontiers in Bioenergy and Biofuels. IntechOpen; 2017. pp. 139-151. DOI: 10.5772/65690

[82] Chabas S. Matériaux biosourcés: le Miscanthus au chevet du bloc porteur en béton. BatiActu, developpement durable. 2017

Chapter 2

Soil Moisture Regime and Mound Position Effects on Soil Water and Vegetation in a Native Tallgrass Prairie in the Mid-Southern United States of America

Tyler J. Durre, Kristofor R. Brye, Lisa S. Wood and Edward E. Gbur

Abstract

Prairie mounds are unique soil surface features that will become increasingly scarce as native tallgrass prairies are continually lost. This study aimed to evaluate (i) whether the soil moisture regime (SMR), mound position, and soil depth affect soil volumetric water content (VWC) and (ii) whether the SMR and mound position affect vegetation over time. Soil VWC was measured continuously from April 2017 to June 2018, and vegetation was sampled in June and August 2017 and in May and August 2018. Maximum VWC for selected rainfall events was ~ 2.5 times greater at 10 cm in the aquic inter-mound than the udic mound position at 30 cm. Soil dry-down rates were four times greater in the udic soil at 10 cm than the aquic soil at 30 cm. Aboveground plant biomass was numerically largest (8489 kg ha^{-1}) at the aquic summit in August 2018 and smallest (1280 kg ha^{-1}) at the aquic inter-mound in May 2018. Results clearly demonstrate the effects that prairie mound topography and differing SMRs have on soil water dynamics and prairie vegetation and suggest that management efforts need to account for mound topography and SMR in order to be most successful.

Keywords: Ozark Highlands, Arkansas, udic, aquic, prairie mounds

1. Introduction

Before the onset of European agriculture, 1.62×10^8 ha of prairie covered the vast area of land from Canada to Mexico and from the Rocky Mountains to western Indiana, known as the Great Plains [1, 2]. Tallgrass prairies once encompassed 6.0×10^7 ha from Canada and Minnesota south to Texas and were the dominant pre-settlement vegetation type in the eastern third of the Great Plains [2, 3]. Since 1830, tallgrass prairie loss in the United States is estimated between 82 and 99%, exceeding the loss of any other major ecosystem in North America [2]. Due to the substantial prairie loss, tallgrass prairies are now considered to be North America's most endangered ecosystem [3]. Factors including conversion to farmland, introduction of non-native forage crops, woody plant encroachment, overgrazing, and urban expansion have contributed to the reduction of tallgrass prairies in North America [2, 4].

Tallgrass prairies are the most mesic prairie variety, and as a result, multiple resources, including soil moisture, may control net primary productivity (NPP) in this ecosystem [5]. Evidence of differing soil moisture dynamics in mounded and inter-mound soil have been described in various field studies [6–8]. Research has generally concluded that inter-mound soils are wetter, often possessing greater water contents, than mounded soil profiles [6–8]. Water content measurements conducted by Ross et al. [8] on a silt loam surface in northwestern Minnesota indicated that mounded soils contained lower water contents at respective depths than the inter-mound soil. Profile descriptions of mounded and inter-mound soils have noted that redoximorphic (redox) concentrations and depletions occur at shallower depths in inter-mound soils, further substantiating that inter-mound soils are wetter than mounded soil [6, 9]. Common depletions were identified in the surface horizon of an inter-mound profile, whereas depletions were absent in the corresponding mounded profile in the top 85 cm in the Arkansas River Valley within the Ouachita physiographic province [9]. Crayfish (*Cambarus* spp.) chimneys are commonly reported in inter-mound soils, but are rarely present in mounded soils, which again suggests that inter-mounds contain more moisture than mounded soils [7, 9]. Additionally, studies have indicated that water is retained longer in inter-mound profiles than in mounded soils [7]. Water is likely retained in the inter-mound for longer periods of time because mounded positions have greater permeability and internal drainage and lower clay contents than inter-mound soils, which increases water movement through the mounded soil profile [7].

The differing water dynamics between mounded and inter-mound soil profiles described in previous studies would likely lead to differences in biomass production and differing plant communities between the mound positions. Studies have characterized herbage production on mounds compared with inter-mounds, differences in vegetation composition (i.e., grass or forb dominated), as well as similarities between plant composition of mounds and inter-mounds [8, 10–12]. Studies conducted by Allgood and Gray [10] on a silt loam surface in eastern Oklahoma and McGinnies [12] on a silt loam soil in Colorado analyzed herbage production of mounds compared with inter-mound mound positions and concluded that mounds generally produce more biomass than inter-mounds. A study conducted by McGinnies [12] in Colorado on a silt loam mounded soil and a loam inter-mound soil noted that the air-dry herbage yields were 94, 180, 323, 358, and 542% greater on seeded mounds than on seeded inter-mounds for intermediate wheatgrass (*Thinopyrum intermedium*), crested wheatgrass [*Agropyron cristatum* (L.) Gaertn.], smooth brome (*Bromus inermis* Leyss.), Russian wildrye [*Psathyrostachys juncea* (Fisch.) Nevski], and big bluegrass (*Poa secunda* J. Presl), respectively.

Studies analyzing whether grasses or forbs were more abundant on mounded positions have yielded mixed results [8, 10]. Scientists have hypothesized that mounds containing pocket gophers (*Geomys bursarius*) will tend to be dominated by grasses, as pocket gophers primarily feed on forb species [10, 13]. Additionally, mound size may determine whether grasses or forbs are the dominant form of vegetation [8]. At the Waubon Prairie in northwestern Minnesota on a silt loam surface, small mounds were generally dominated by grasses, whereas medium-sized mounds were forb-dominant and large mounds were comprised mostly of shrubs [8]. Additionally, vegetation differences between mounds and the surrounding prairie occur because mounded soils exhibit increased biological soil disturbance compared with inter-mound soils [11]. As soil is continually disturbed, vegetation succession occurs, which promotes the abundance of pioneer forb species and other disturbance-tolerant plants [8, 11].

Studies determining whether plant species richness was greater on mound or inter-mound mound positions have also provided mixed results. Brotherson [11] concluded that the species richness on the mound was only slightly larger than the species richness of the corresponding inter-mound in Iowa, with 51 plant species identified on the mounds and 48 species identified in the inter-mound on soils with loam, clay loam, and silty clay loam soil textures [11, 14]. Of the 51 plant species present on the mounds, 38 were also present in the adjacent prairie [11]. Conversely, Allgood and Gray [10] noted that 18 plant species were identified on inter-mound soils, whereas 13 plant species were identified on mounded soils on a silt loam soil in eastern Oklahoma. Of the 18 species located on inter-mound soils, 6 were also located on mounded soils [10]. Although the studies may disagree on whether species richness was greater in mound or inter-mound soils, both studies demonstrated that a degree of dissimilarity between plant species comprising mounds and inter-mounds exists. Scientists have hypothesized the reason for the dissimilarity between mounds and inter-mounds is due to the microtopographic variation of the mounds compared with inter-mound soils [11, 15].

Studies analyzing soil moisture with time and vegetation in tallgrass prairies within the Ozark Highlands are of interest as the Ozark Highlands occupies a topographic, climatic, and botanical transition zone from the grassland-dominated Great Plains to the west and northwest to the warm and wetter forest to the east and southeast [1, 16, 17]. The Ozark Highlands Major Land Resource Area (MLRA) 116A occupies portions of eastern Oklahoma, northwestern and north-central Arkansas, and southwestern to south-central Missouri and is approximately 85,720 km^2 [18]. The Ozark Highlands land cover distribution is characterized as approximately 54% forest, 33% grasslands, 5% cropland, 4% urban development, 3% water, and 1% other [19]. The forested region of the Ozark Highlands is inhabited by oak (*Quercus* spp.), hickory (*Carya* spp.), and shortleaf pine (*Pinus echinata* Mill.) [19]. Common grassland species present in the Ozark Highlands include fescue (*Festuca* L.), big bluestem (*Andropogon gerardii* V.), little bluestem [*Schizachyrium scoparium* (Michx.) Nash], indiangrass [*Sorghastrum nutans* (L.) Nash], and dropseeds (*Sporobolus* spp.) [19]. Corn (*Zea mays* L.) and soybean [*Glycine max* (L.) Merr.] are agronomic crops typically grown in the Ozark Highlands.

Alfisols and Ultisols are the dominant soil orders present in the Ozark Highlands [20]. Limestone, dolomite, and occasionally sandstone are common parent materials in the region [20]. Argillic horizons have developed over time as physical and chemical weathering has caused the cherty limestone parent material to disintegrate into chert and clay [20]. Soils in the Ozark Highlands are shallow to very deep, moderately to excessively well-drained, and medium- to fine-textured [20].

Prairie mounds have been described as soils with special scientific value, but relatively little research has been conducted on undisturbed prairie mounds in native tallgrass prairies [21]. Most research of prairie mounds has occurred on the west coast [15, 22, 23], but few studies have been performed in the mid-southern region of the United States. Additionally, most prairie mound research has focused on determining valid hypotheses for mound formation, while various aspects of prairie mounds have been studied specifically in northwest Arkansas [24, 25], in northeast Arkansas [26], and in central and southern Arkansas [9, 27]. Though various studies have reported soil moisture differences between mound and inter-mound areas, none of the studies evaluated soil moisture dynamics over extended time periods and multiple seasons. In additional, potential vegetation differences in mounded ecosystems in Arkansas have not been researched. Therefore, the objective of this field study was twofold: (i) characterize soil volumetric water content (VWC) differences between landscape positions (i.e., mound summit and inter-mound) over time and among soil depths (i.e., 10, 20, 30, and 50 cm) in contrasting soil moisture regimes (SMR) (i.e., aquic and udic) and (ii) determine the effect of landscape position (i.e., mound and inter-mound), soil moisture regime (i.e., aquic and udic), and time on vegetative properties [i.e., total productivity, total diversity, species evenness, species richness, vegetation similarity, and grass abundance compared with other species abundance (i.e., sedges, rushes, and forbs)] in a native tallgrass prairie in the Ozark Highlands region of northwest Arkansas. It was hypothesized that numerous differences in soil moisture, vegetation, and soil morphology would exist with depth among the various mound positions across soil moisture regimes.

2. Materials and methods

2.1 Site description

Research for this field study began in April 2017 at the Chesney Prairie Natural Area, hereafter referred to as Chesney Prairie, located near Siloam Springs, Benton County, Arkansas (36°13′12″ N lat., 94°28′57″ W long., **Figure 1**). Chesney Prairie is part of the Ozark Highlands (MLRA 116A) [18].

The Chesney Prairie (**Figure 1**) is a tallgrass prairie that has been managed by the Arkansas Natural Heritage Commission (ANHC) since 2000 [28]. Chesney Prairie is a 33-ha remnant of prairie ecosystems that formerly encompassed over 30,000 ha of the Ozark Plateau and is one of the few prairie remnants on the Arkansas portion of the Springfield Plateau [29]. In addition, Chesney Prairie and the nearby Stump Prairie are the two remaining native prairie remnants of Lindsley's Prairie, which once encompassed approximately 6200 ha around present-day Siloam Springs, AR [30].

Chesney Prairie is a diverse prairie that supports over 450 plant species, including 290 native plant species and 18 rare plant species [29]. Big bluestem (*Andropogon gerardii* V.), little bluestem [*Schizachyrium scoparium* (Michx.) Nash], indiangrass [*Sorghastrum nutans* (L.) Nash], and switchgrass (*Panicum virgatum*) are typical prairie grasses present at Chesney Prairie [30]. Common forb species inhabiting Chesney Prairie include large flower tickseed (*Coreopsis grandiflora*), prairie

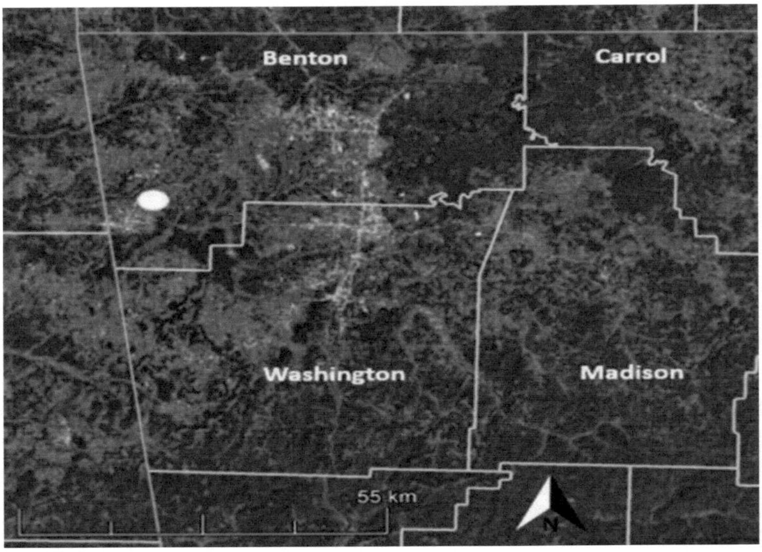

Figure 1.
Map depicting the approximate location of Chesney prairie (represented with oval) located in Benton County, Arkansas.

grayfeather (*Liatris pycnostachya*), and rattlesnake master (*Eryngium yuccifolium*). Periodic prescribed burns and invasive species eradication are management practices currently used to increase the native plant population [30]. Prescribed burning has occurred approximately every 3 years, with the last burn occurring in January 2017.

Chesney Prairie contains two soil series: Jay silt loam (fine-silty, mixed, active, thermic Oxyaquic Fragiudalf), which is in a udic soil moisture regime, and Taloka silt loam (fine, mixed, active, thermic Mollic Albaqualfs), which is in an aquic soil moisture regime [31, 32]. The macroscale slope is approximately 4%, and the land surface undulates some throughout the entire Chesney Prairie area. However, slopes are ≤2% within each soil mapping unit [1]. Numerous prairie mounds are present at Chesney Prairie, and the prairie is divided by Sager Creek, an ephemeral stream. The prairie mounds are ~20.9 m in diameter and ~0.7 m in height and are roughly circular.

The mean average air temperature throughout the region containing the Chesney Prairie over the past 30 years was 14.9°C, with an average January minimum of 2.9°C and an average July maximum of 26.1°C [33]. The mean annual precipitation over the past 30 years was 1203 mm, with approximately 64% of the rainfall occurring during the growing season from April to October [33].

2.2 Soil water content monitoring

To continuously monitor changes in soil VWC with depth over time, two prominent mounds were identified in both the Jay and Taloka soil series, and the distance from summit to summit was measured. The inter-mound position, defined as the midpoint between the mound summits, was marked. On 8 April 2017, at both the inter-mound positions between the two mound summits and at one of the adjacent mound summits in both soil series, a small trench was manually excavated after cutting and removing the top layer of sod. Water content reflectometers (model CS615, Campbell Scientific, Inc., Logan, UT) were installed horizontally at depths of 10, 20, 30, and 50 cm below the soil surface. The small trench was filled back in with soil from the appropriate natural horizon, and the intact piece of sod was placed back on top where it was removed from to maintain a minimally disturbed appearance (**Figure 2**). The water content reflectometer wires were shallowly buried and connected to a datalogger (model CR10X, Campbell Scientific, Inc.) to record data every 5 minutes and output mean volumetric soil water contents hourly. Approximately weekly, data were manually transferred to a storage module (model SM16M, Campbell Scientific, Inc.) using a keyboard display (model CR10KD, Campbell Scientific, Inc.) and transferred to a desktop computer. Volumetric soil water contents were measured and recorded through 30 June 2018.

To determine the effects of mound position (i.e., mound summit and inter-mound), depth below the soil surface (i.e., 10, 20, 30, and 50 cm), and soil moisture regime (i.e., aquic and udic) on soil volumetric water content dynamics, dry-down periods were determined for each major rainfall event between 1 June 2017 and 31 May 2018. Dry-down periods for each depth were identified as the linear phase between the maximum and minimum soil water content measured for each event before the next wetting event occurred. The maximum and minimum soil water contents for each depth were also recorded for each rainfall event for subsequent analyses. Water content maxima and the soil water content 2 days after the maximum was achieved were used to calculate the rate of dry-down for selected rainfall events.

2.3 Weather station

A micrometeorological weather station was erected on-site on 15 April 2017 in the Jay soil series area at Chesney Prairie to measure rainfall, air temperature, and

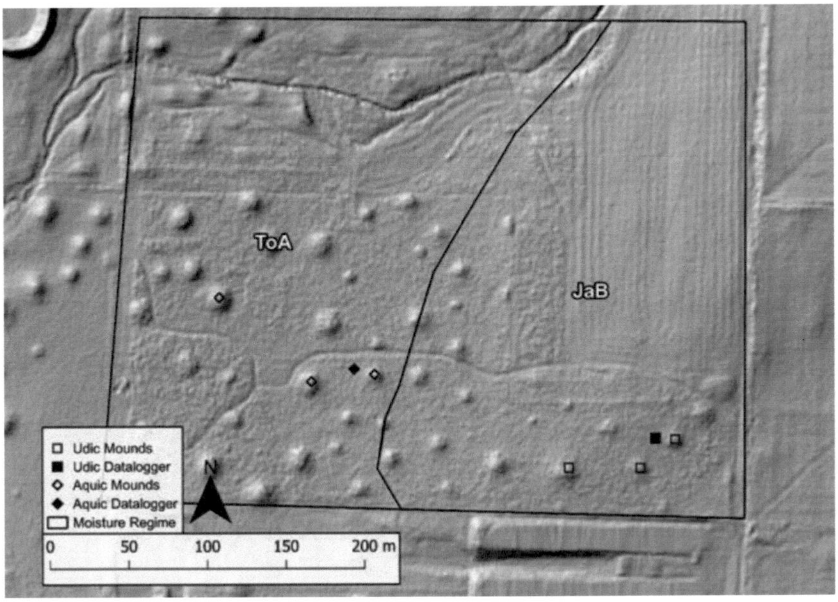

Figure 2.
Satellite imagery depicting the locations of the aquic and udic volumetric water content dataloggers and all mounds sampled within the aquic (i.e., ToA) and udic (i.e., JaB) soil moisture regimes at Chesney prairie. Data downloaded from Arkansas GIS Office [48].

relative humidity. The weather station contained a 25-cm-diameter tipping bucket rain gauge (model TR-525 M, Texas Electronics, Inc., Dallas, TX) and a combined air temperature/relative humidity sensor (model HMP50, Campbell Scientific, Inc., Logan, UT). Both sensors were connected to a datalogger (model CD10X, Campbell Scientific, Inc.), which recorded data every 2 minutes and output data summaries every hour. Approximately weekly, data were manually collected on a storage module (model SM16M, Campbell Scientific, Inc.) using a keyboard display (model CR10KD, Campbell Scientific, Inc.) and transferred to a desktop computer. Precipitation, air temperature, and relative humidity were measured and recorded through 30 June 2018.

2.4 Vegetation sampling and analysis

Vegetation samples were collected on 2–3 June and 17–18 August 2017 and 19 May and 16 August 2018 from mound summit and inter-mound positions in the Jay and Taloka soil series. At each position, all vegetation within a 0.25-m^2 metal frame was cut to approximately a height of 2 cm. Stem by stem, the cut vegetation was bagged separately as either a grass or other (i.e., a sedge, rush, shrub, etc.). In total, three vegetation samples were collected at mound summit and inter-mound positions in each soil series on each sample date. Vegetation samples were oven dried at 55°C for at least 5 days and weighed to determine dry matter by vegetation type (i.e., grasses or other).

Dry matter data in May 2018 were used to determine vegetation diversity using the Shannon-Wiener index [34]. Each plant species within the 0.25 m^2 metal frame was identified to determine the species richness for the site. The number of a given plant species was recorded and divided by the total number of plants observed to calculate the relative abundance for each species. The relative abundance of each plant species was used in the Shannon-Wiener equation to calculate the diversity index. The resulting diversity index and species richness were then used to calculate evenness. The Shannon-Wiener and evenness equations are outlined below:

$$H = -\sum_{i=1}^{n}(pi)\ \ln(pi) \tag{1}$$

where H is the Shannon-Wiener Index, s is the number of species, and pi is the proportion of total sample belonging to the ith species, and

$$\mathrm{EH} = \mathrm{H}/\ln(\mathrm{s}) \tag{2}$$

where EH is evenness, H is the Shannon-Wiener Index, and s is the number of species. Additionally, a Sorenson coefficient was calculated using Eq. (3) to determine the similarity of vegetation comprising the mounded and inter-mound positions within and across soil moisture regimes:

$$\mathrm{Ss} = 2\mathrm{a}/(2\mathrm{a} + \mathrm{b} + \mathrm{c}) \tag{3}$$

where a is the number of species both locations have in common, b is the number of species present in only location one, and c is the number of species present in only location two.

2.5 Statistical analysis

Based on a completely random design, an analysis of variance (ANOVA) was conducted using PROC MIXED in SAS 9.4 to evaluate the effects of soil moisture regime (i.e., udic and aquic), mound position within soil moisture regime (i.e., mound summit and inter-mound within the aquic and udic soil moisture regimes), time (i.e., wet and dry season), depth (i.e., 10, 20, 30, and 50 cm) and their interaction on soil water content maxima and minima achieved, and the rate of dry-down during drying events. Multiple drying events isolated over time served as temporal replication for these analyses.

Based on a split-split plot, completely random experimental design, a three-factor ANOVA was conducted using PROC MIXED in SAS 9.4 to evaluate the effects of soil moisture regime (i.e., aquic and udic), mound position (i.e., mound summit and inter-mound), time (i.e., sample date), and their interactions on aboveground dry matter production. The whole-plot factor was soil moisture regime, the split-plot factor was mound position, and the split-split-factor was time. A four-factor ANOVA was conducted in SAS 9.4 to evaluate the effects of soil moisture regime (i.e., aquic and udic), mound position (i.e., mound summit and inter-mound), time (i.e., sampling date), biomass type (i.e., grasses or other species), and their interactions on total dry matter production. Lastly, a two-factor ANOVA was conducted in SAS 9.4 to determine the effects of soil moisture regime (i.e., aquic and udic), position (i.e., mound summit and inter-mound), and their interaction on Shannon-Wiener diversity and species richness and evenness. For all analyses, the least significant difference (LSD) was used to separate means at the 0.05 level.

3. Results and discussion

3.1 Water content dynamics

Soil water contents exhibited distinct trends with time at both mound positions (i.e., summit and inter-mound) within both soil moisture regimes (i.e., aquic and udic). Precipitation totaled 117.5 cm at the field site from 1 June 2017 to 31 May 2018 and was within 10% of the 30-year normal annual precipitation (120.3 cm) for the region, designating the current year as a typical/average year for the region encompassing the study site (**Figure 3**). In total, 112 independent precipitation events (i.e., periods of precipitation of any magnitude separated by half a day without precipitation) occurred from mid-April 2017 to 31 May 2018. Of the 112 precipitation events, 95 occurred during 1 June 2017 to 31 May 2018. Approximately 59% of the precipitation events within the study period caused a clear response (i.e., a response that could be easily differentiated from normal fluctuations in VWC) in the 10-cm sensor for each mound position within both soil moisture regimes, while only 14 of the 95 precipitation events caused a clear response in all 16 sensors (**Figures 4** and **5**).

Seasonal wet-up and dry-down trends resulting from precipitation patterns were evident at each mound position within both soil moisture regimes and were most pronounced at the aquic inter-mound (**Figures 4** and **5**). Seasonal dry-down periods began in early summer [approximately day of year (DOY) 170] with the subsequent wet-up period beginning in late fall (approximately DOY 300), continuing through spring (**Figures 4** and **5**). Noticeable wet-up and dry-down periods have been recorded in previous research by Briggs and Knapp [5], who observed seasonal dry-down periods beginning in late summer and wet-up periods occurring in spring and early summer at depths of 25 and 100 cm over an 11-year period at the Konza Prairie. Additionally, the annual soil volumetric water content (VWC) fluctuations in the current study roughly followed the four phases of annual soil moisture as described by Illston et al. [35]. In Oklahoma, the statewide soil fractional water index (FWI) entered a moist plateau phase from November to mid-March, a transitional drying phase from mid-March to mid-June, an enhanced drying phase from mid-June to late August, and ending with the recharge phase from late August to November [35]. In the current study, the moist plateau period occurred between mid-February and May (**Figures 4** and **5**). During the moist plateau phase, volumetric water contents were at their largest and were relatively consistent due to reduced evaporation and evapotranspiration from low sun angles and dormant vegetation [35]. The transitional drying phase, characterized by a gradual decrease in VWC from increased evapotranspiration from growing vegetation [35], occurred from June to early July, followed by the enhanced drying stage from early July to early October. During the enhanced drying stage, soil VWCs decline sharply to their seasonal low due to continued evapotranspiration and limited inputs of water from

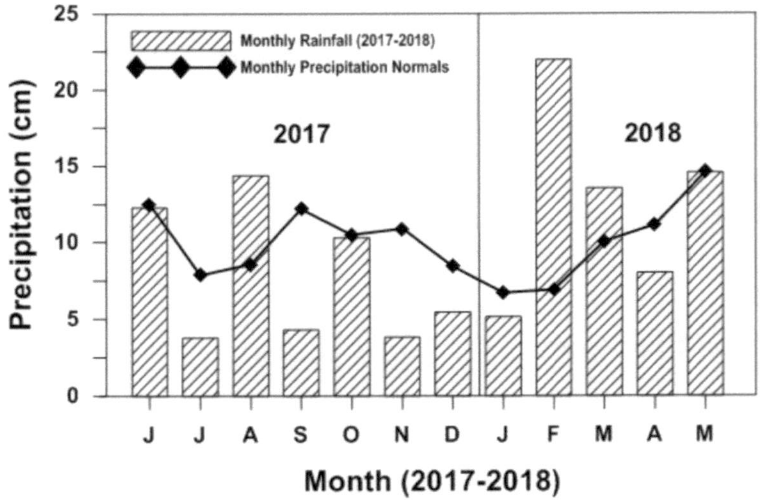

Figure 3.
Monthly precipitation recorded at the study site compared to the 30-year normal monthly precipitation for the region encompassing the study site from June 2017 to May 2018.

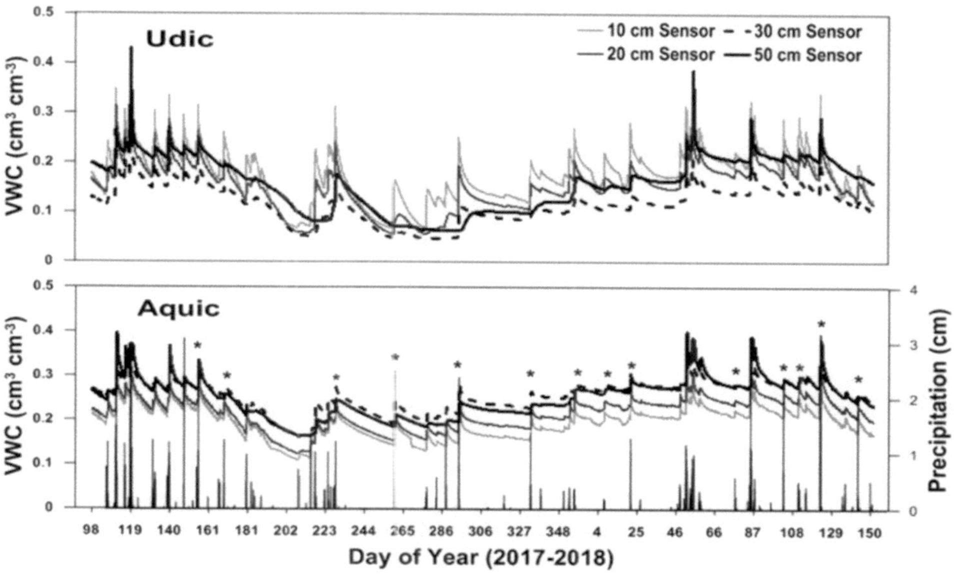

Figure 4.
Volumetric water content and precipitation over time with depth (10, 20, 30, and 50 cm) for the udic and aquic mound summit. Precipitation events denoted with an asterisk were used for statistical analysis.

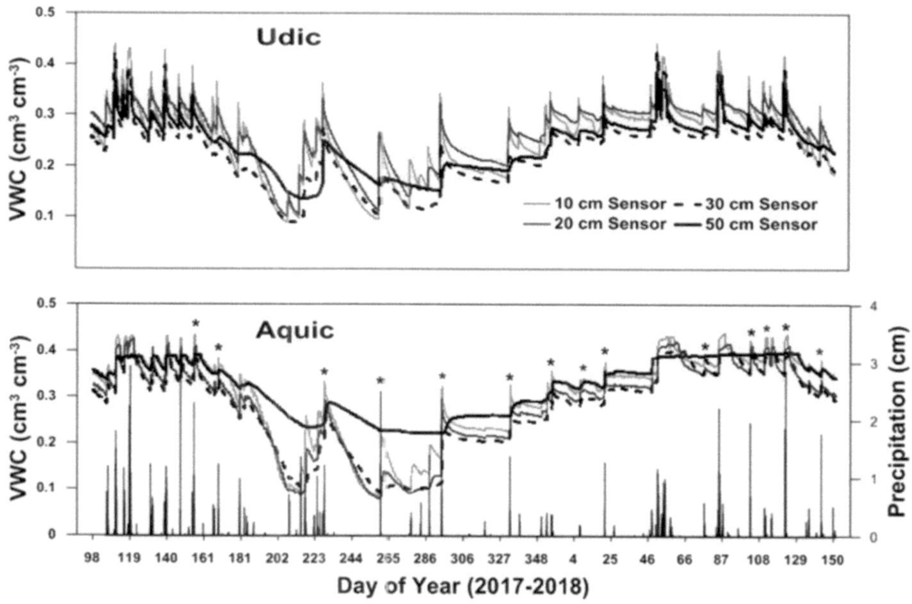

Figure 5.
Volumetric water content and precipitation over time with depth (10, 20, 30, and 50 cm) for the udic and aquic inter-mound. Precipitation events denoted with an asterisk were used for statistical analysis.

precipitation [35]. Lastly, the soil VWC gradually increased from early October to early February during the recharge phase, as a result of decreased evapotranspiration due to low sun angles and inputs of water from precipitation [35].

In general, the mound positions within the aquic soil had larger VWCs over time at respective depths than the corresponding mound position in the udic soil moisture regime (**Figures 4** and 5). Additionally, the inter-mound positions generally contained larger VWCs at respective depths than the mound summit of the same soil moisture regime, as expected based on the soil morphological characteristics (**Figures 4** and **5**). The results of the study agree with observations [6, 7, 9] and measurements [8] of water content in mounded and inter-mound profiles from past research. In the Arkansas River Valley, in a silt loam surface, redox depletions were identified at the soil surface in inter-mound pedons, whereas depletions were not present in mounded soil profiles until a depth of 85 cm, indicating that inter-mound soils are generally wetter than mounded profiles [9]. Soil moisture was likely greater and retained longer in inter-mound soil profiles due to greater clay concentrations and lower saturated hydraulic conductivities typical of inter-mound soils than in mounded profiles [7, 36].

Volumetric water contents in the udic mound were generally largest at the 10-cm depth and lowest at the 30-cm depth, whereas VWCs were generally largest at either the 30- or 50-cm depths and smallest at the 10-cm depth in the aquic mound (**Figure 4**). Additionally, seasonal dry-down was more pronounced in the udic mound, in which VWCs at all depths fell below 0.1 cm^3 cm^{-3}, than the aquic mound, which recorded no VWCs lower than 0.1 cm^3 cm^{-3} (**Figure 4**). Volumetric water contents in the udic inter-mound were generally largest at either the 10- or 20-cm depth and lowest at 30 cm, whereas VWCs were generally largest at the 10-cm depth during wet-up periods and at the 50-cm depth during periods of dry-down in the aquic inter-mound positions (**Figure 5**). As with the udic inter-mound, VWCs were generally lowest at 30 cm in the aquic inter-mound (**Figure 5**). The seasonal dry-down period was more pronounced in the aquic inter-mound than in the udic inter-mound, with exception of the 50-cm depth in the aquic inter-mound, which was not impacted by dry-down as dramatically as the 10-, 20-, and 30-cm depths (**Figure 5**).

The magnitude and frequency of response to precipitation events appeared to be larger for the surface sensors (i.e., 10 and 20 cm) than for the 30- and 50-cm sensors for each mound position (**Figures 4** and **5**). Additionally, the magnitude of response to a precipitation event was generally larger in the udic soil moisture regime than in the aquic soil moisture regime when similar mound positions were compared (**Figures 4** and **5**). Similar soil water content trends were noted in Briggs and Knapp [5], in which larger and more numerous maxima were observed over time in the 25-cm sensor than in the 100-cm sensor, indicating that soil near the surface was more influenced by wet-up and dry-down events than soil deeper in the profile. Surface soil layers likely exhibited larger decreases in VWC during dry-down events than subsurface layers due to losses of water through evapotranspiration and/or vertical drainage. Additionally, surface sensors likely responded to rainfall events more frequently than subsurface sensors due to redistribution of water in the soil profile. Most of the water that infiltrates into the soil surface from a precipitation event will likely percolate through the surface soil layers (i.e., 10 and 20 cm). However, the amount of water reaching the subsurface (i.e., 30 and 50 cm) soil layers may be diminished as water is extracted by plants, which would then require a larger precipitation event to occur before water contents at lower soil depths increase. In addition, subsurface soils may respond to fewer precipitation events because they are more influenced by additions of water from deeper in the soil profile (i.e., a seasonal high water table) as opposed to additions of water from the soil surface. The effect of a seasonal high water table on soil volumetric water content was clearly demonstrated at the 50-cm depth in the aquic inter-mound from approximately DOY 46 to 130 (**Figure 5**). From DOY 46 to 130, soil water contents in the 10-, 20-, and 30-cm depth fluctuated from multiple wet-up and dry-down events, whereas the 50-cm depth gradually increased with no distinct peaks, indicating that the 50-cm depth was more influenced by water moving upwards from deeper in the soil profile than from water moving downward from precipitation events.

3.2 Soil hydraulic properties

All soil hydraulic properties were affected by one or more or a combination of treatment factors evaluated (i.e., SMR, mound position within SMR, season, and/or soil depth). Maximum and minimum soil VWCs differed ($P < 0.05$) by depth within respective mound positions across SMRs (**Table 1**). Maximum soil VWC was numerically largest (0.39 cm^3 cm^{-3}) at the 10-cm depth in the aquic inter-mound and was significantly smallest (0.16 cm^3 cm^{-3}) at the 30-cm depth in the udic mound (**Figure 6**). Additionally, maximum soil VWCs were at least numerically greater in the aquic and udic inter-mound positions than in the corresponding mound summits at each depth (**Figure 6**). When respective mound positions were compared across SMRs, the maximum VWC was at least numerically greater in the aquic mound position than that of the udic mound position for a given depth, excluding the mound summit at the 10-cm depth (**Figure 6**).

The mean minimum soil VWC was numerically largest (0.29 cm^3 cm^{-3}) at the 50-cm depth in the aquic inter-mound and numerically smallest (0.11 cm^3 cm^{-3}) at the 30-cm depth in the udic mound (**Figure 6**). Additionally, the minimum soil VWC was at least numerically larger at respective depths at the inter-mound position than at the mound summit within the same SMR (**Figure 6**). When respective mound positions were compared across SMRs, the minimum VWC was at least numerically greater in the aquic than that of the udic mound position for a given depth (**Figure 6**).

Averaged over mound position within SMR and season, both maximum and minimum soil VWCs differed ($P < 0.05$) with depth across SMRs (**Table 1**). Mean maximum VWC was numerically largest (0.32 cm^3 cm^{-3}) at the 10-cm depth in the udic SMR and smallest (0.23 cm^3 cm^{-3}) in the udic SMR at the 30-cm depth. The maximum VWC was greater at each depth interval in the aquic than udic SMR, with exception of at the 10-cm depth, in which maximum VWC in the SMRs did

Source of variation	Max VWC	Min VWC	DDR
		P	
Soil moisture regime	0.001	0.003	0.016
Position within SMR [Pos(SMR)]	<0.001	0.005	0.714
Season (S)	0.959	0.062	0.355
Depth (D)	<0.001	0.001	<0.001
SMR × S	0.299	0.419	0.424
Pos × S(SMR)	0.667	0.310	0.852
SMR × D	<0.001	0.004	<0.001
SMR × D(Pos)	<0.001	0.001	0.122
S × D	0.003	0.561	0.014
SMR × S × D	0.765	0.142	0.850
Pos × S × D(SMR)	0.933	0.854	0.906

Table 1.
Summary of the combined effects of soil moisture regime (aquic and udic) position within soil moisture regime (summit or inter-mound within the aquic and udic soil moisture regime), season (wet and dry), and depth (10, 20, 30, and 50 cm) on soil maximum volumetric water content after a rainfall event (max VWC), minimum volumetric water content after a rainfall event (min VWC), dry-down rate (DDR), and lag time (LT) in a mounded native tallgrass prairie in the Ozark highlands region of Northwest Arkansas.

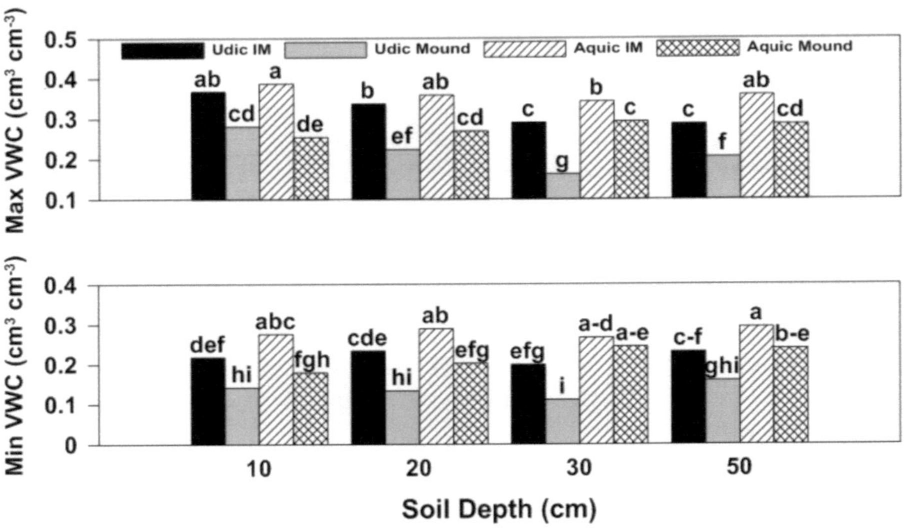

Figure 6.
The effects of soil depth averaged over mound position and soil moisture regime on maximum volumetric water content (Max VWC) and minimum volumetric water content (Min VWC) for selected precipitation events in aquic and udic soils in a mounded native tallgrass prairie in the Ozark highlands region of Northwest Arkansas. Means with different letters are significantly different at the 0.05 level.

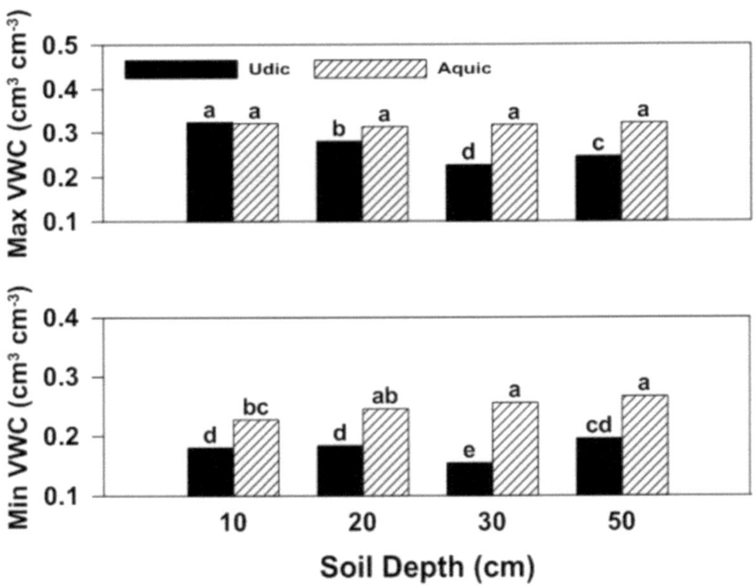

Figure 7.
Soil depth effects on maximum volumetric water content (max VWC) and minimum volumetric water content (Min VWC) for selected precipitation events in aquic and udic soils in a native tallgrass prairie containing prairie mounds in the Ozark highlands region of Northwest Arkansas. Means for a soil property with different letters are significantly different at the 0.05 level.

not vary. Minimum soil VWC was numerically largest (0.27 cm^3 cm^{-3}) at the 50-cm depth in the aquic and smallest (0.16 cm^3 cm^{-3}) at the 30-cm depth in the udic SMR (**Figure 7**). Compared across SMRs, the aquic soil contained a minimum VWC that was, on average, 1.4 times larger than that of the udic soil (**Figure 7**).

Averaged over mound position, SMR, and position within SMR, maximum VWC differed (P = 0.003) between seasons by depth (**Table 1**). Maximum soil VWC was largest (0.33 cm^3 cm^{-3}) during the dry season at 10 cm and numerically smallest (0.27 cm^3 cm^{-3}) during the dry season at 30 cm (**Figure 8**). When seasons were compared, maximum VWC was larger in the dry season than wet season at 10 cm, larger in the wet season than the dry season at 50 cm, and did not vary by season at depths of 20 and 30 cm (**Figure 8**). Additionally, averaged over SMR, depth, and season, maximum and minimum VWCs differed ($P < 0.05$) between mound positions within SMRs (**Table 1**). Maximum VWC was largest (0.36 cm^3 cm^{-3}) in the aquic inter-mound and smallest (0.22 cm^3 cm^{-3}) in the udic summit (**Figure 9**). The aquic soil contained larger maximum VWCs at each mound position (**Figure 9**). Similar to the maximum VWCs, minimum VWC was largest (0.28 cm^3 cm^{-3}) at the aquic inter-mound and smallest (0.14 cm^3 cm^{-3}) at the udic summit (**Figure 9**). The aquic SMR contained a larger minimum VWC than the udic soil when respective mound positions were compared (**Figure 9**).

The aquic soil likely had larger maximum and minimum VWCs than the udic soil based on characteristics of the two soil series. The internal drainage of the

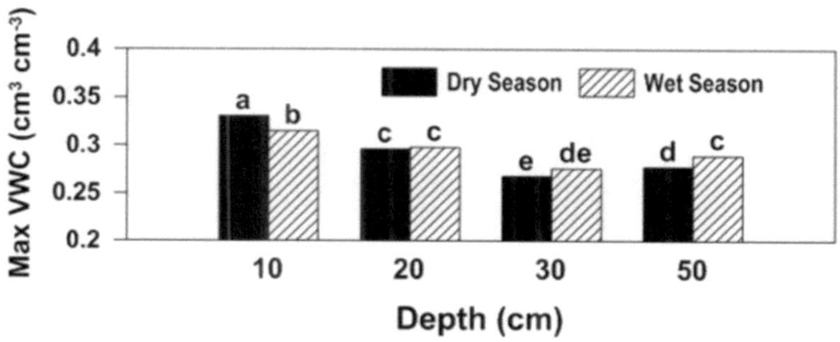

Figure 8.
Soil depth effects on maximum volumetric water content for selected precipitation events during the wet and dry season in a native tallgrass prairie containing prairie mounds in the Ozark highlands region of Northwest Arkansas. Means with different letters are significantly different at the 0.05 level.

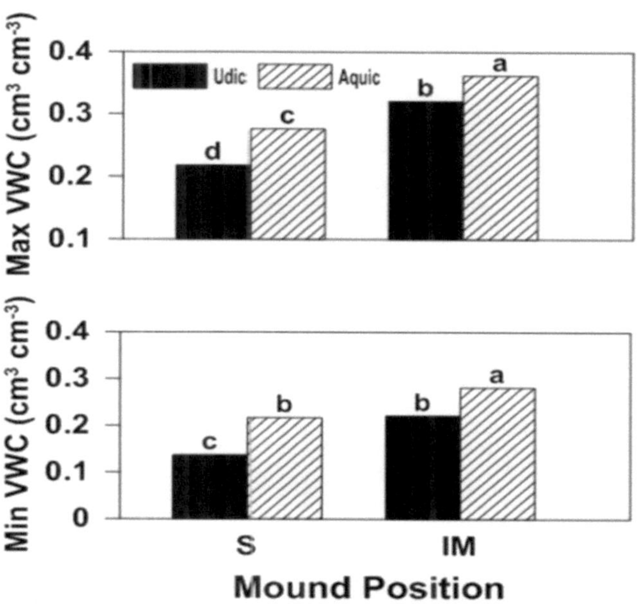

Figure 9.
Mound position within soil moisture regime effects on maximum volumetric water content (Max VWC) and minimum volumetric water contents (Min VWC) for selected precipitation events in native tallgrass prairie with prairie mounds in the Ozark highlands region of Northwest Arkansas. Means for a soil property with different letters are significantly different at the 0.05 level.

aquic SMR (i.e., Taloka soil series) is characterized as somewhat poorly drained, which would retain more water than the moderately well-drained Jay soil series (i.e., udic SMR). Similar trends in soil moisture were noted by Henninger et al. [37] in east-central Pennsylvania across six soil series [i.e., Klinesville (Lithic Dystrudepts), Calvin (Typic Dystrudepts), Leck Kill (Typic Hapludults), Hartleton (Typic Hapludults), Albrights (Aquic Fragiudalfs), and Alvira (Aeric Fragiaquults)]. The somewhat poorly drained Alvira soil maintained greater soil moisture levels than the moderately well- to well-drained soils for the duration of the study [37]. Differences in maximum and minimum VWCs between the mound summit and inter-mound mound positions likely resulted from differing clay concentrations, soil organic matter (SOM) contents, and estimated bulk densities. Inter-mound clay concentrations were at least numerically larger than that in the corresponding mound summit position at each 10-cm depth interval to a depth of 90 cm in both SMRs (**Table 2**). Increased clay concentrations would result in greater water-holding capacity, accounting for the greater maximum and minimum VWCs in inter-mound mound positions. Additionally, larger maximum and minimum VWCs in the inter-mound position may be attributed to greater SOM. Soil organic matter has the ability to absorb water and promote soil aggregation, both of which enhance soil water-holding capacity. According to Scott et al. [38], for every 1% of SOM, the soil can hold 154,340 liters of plant-available water per hectare to a depth of 1 m. In the current study, SOM contents were at least numerically larger

Soil property	Depth (cm)	Udic		Aquic	
		Summit	Inter-mound	Summit	Inter-mound
Clay (g g^{-1})	0–10	0.05 t[†]	0.06 st	0.07 rst	0.09 p-t
	10–20	0.06 st	0.07 rst	0.08 q-t	0.12 n-r
	20–30	0.06 st	0.11 o-s	0.10 p–t	0.17 k-n
	30–40	0.07 rst	0.18 j-m	0.09 p–t	0.20 h-k
	40–50	0.08 q-t	0.25 d-h	0.10 p–t	0.23 f-j
	50–60	0.09 p-t	0.27 b-f	0.11 o-s	0.26 c-g
	60–70	0.10 p–t	0.30 a-d	0.10 p–t	0.27 b-f
	70–80	0.12 n-r	0.33 a	0.13 m-q	0.31 abc
	80–90	0.13 m-q	0.30 a-d	0.14 l-p	0.32 ab
SOM (Mg ha^{-1})	0–10	50.3 cd	59.7 a	43.1 efg	57.1 ab
	10–20	38.2 g-l	47.7 de	37.2 i-m	39.0 g-k
	20–30	34.7 k-o	41.6 f-i	33.7 l-p	32.5 m-r
	30–40	35.4 j-n	36.6 i-m	33.6 l-p	28.3 q-x
	40–50	30.7 n-t	29.7 o-u	29.9 o-u	24.7 u-B
	50–60	28.3 q-x	25.5 t-A	27.5 r-x	23.3 x-C
	60–70	28.3 q-x	29.0 p-w	29.0 o-w	23.4 x-C
	70–80	23.3 x-C	28.3 q-x	23.8 w-B	20.2 BCD
	80–90	21.4 ABC	27.6 r-x	21.5 ABC	20.1 BCD

[†]*All means for a soil property followed by different letters are significantly different at the 0.05 level.*
Adapted from Durre et al. [36].

Table 2.
Summary of the combined effects of soil moisture regime (aquic and udic), mound position (summit and inter-mound), and soil depth (0–90-cm in 10-cm intervals) on soil clay concentrations and soil organic matter in a native tallgrass prairie with prairie mounds in the Ozark highlands region of Northwest Arkansas.

in inter-mounds to a depth of 20 and 40 cm for the aquic and udic soil, respectively, which would result in a larger water-holding capacity for the inter-mounds than for the mounds at those depths (**Table 2**). Soil bulk density was at least numerically larger at each depth interval in the mound summit, which would account for the lower maximum and minimum VWCs than in the inter-mound position (**Table 3**). Increased bulk densities result in a lower soil water-holding capacity due to decreased total porosity.

Averaged across mound position, SMR, and position within SMR, the rate of dry-down differed ($P = 0.01$) by season with depth (**Table 1**). The soil dry-down rate was greatest (0.029 $cm^3\ cm^{-3}\ day^{-1}$) during the dry season at 10 cm and numerically lowest (0.009 $cm^3\ cm^{-3}\ day^{-1}$) during the wet season at 30 cm (**Figure 10**). Dry-down rates were greater during the dry season than the wet season at 10 cm, but no seasonal differences occurred at the 20- or 30-cm depth (**Figure 10**). Averaged across mound position within SMR and season, soil dry-down rate differed ($P < 0.01$) with depth between SMRs (**Table 1**). Soil dry-down rates were largest (0.032 $cm^3\ cm^{-3}\ day^{-1}$) in the udic at 10 cm and numerically smallest (0.008 $cm^3\ cm^{-3}\ day^{-1}$) in the aquic SMR at 30 cm (**Figure 10**). Though soil dry-down rates were larger in the udic than the aquic SMR at 10 cm, no differences in dry-down rate between SMR were noted at the 20- and 30-cm depths (**Figure 10**).

Soil dry-down rates were likely larger during the dry season at 10 cm due to evapotranspiration. Water added to the soil during the dry season will likely be quickly removed by growing plants, which would increase the rate of soil dry-down.

Soil property	Depth (cm)	Summit	Inter-mound
BD ($g\ cm^{-3}$)	0–10	1.27 k†	1.14 m
	10–20	1.37 gh	1.30 jk
	20–30	1.40 efg	1.35 hi
	30–40	1.40 efg	1.37 gh
	40–50	1.43 cde	1.39 fg
	50–60	1.45 abc	1.39 fg
	60–70	1.44 bcd	1.38 fgh
	70–80	1.47 ab	1.38 fgh
	80–90	1.48 a	1.40 efg
K_{sat} ($mm\ hr.^{-1}$)	0–10	43.7 c	63.4 a
	10–20	31.1 def	32.5 de
	20–30	26.2 fgh	20.1 ijk
	30–40	25.0 ghi	12.2 l-q
	40–50	19.9 jk	7.0 p–t
	50–60	17.5 j-m	6.1 rst
	60–70	17.8 jkl	5.9 st
	70–80	13.3 l-o	4.6 t
	80–90	11.8 m-s	4.9 t

†All means for a soil property followed by different letters are significantly different at the 0.05 level.
Adapted from Durre et al. [36].

Table 3.
Summary of the combined effects of mound position (summit and inter-mound) and depth (0–90-cm in 10-cm intervals) on estimated bulk density (BD) and estimated saturated hydraulic conductivity (K_{sat}) in a native tallgrass prairie with prairie mounds in the Ozark highlands region of Northwest Arkansas.

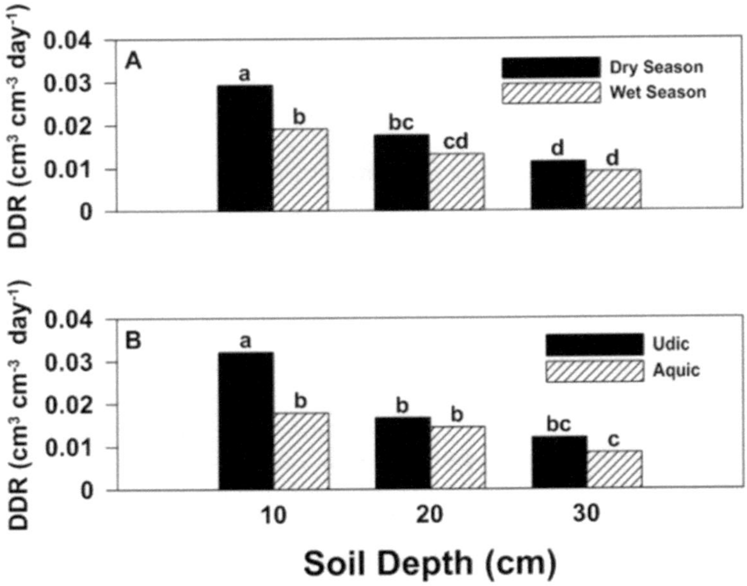

Figure 10.
Soil depth effects on dry-down rates of selected precipitation events during the (A) wet and dry season and in (B) udic and aquic soil moisture regimes in a native tallgrass prairie containing prairie mounds in the Ozark highlands region of Northwest Arkansas. Means with different letters are significantly different at the 0.05 level.

A study conducted by Henninger et al. [37] in east-central Pennsylvania noted that water entering the top 15 cm of soil during the summer months was quickly depleted by evapotranspiration resulting in annually low soil moisture contents in each soil series studied. As evapotranspiration decreased in September, the soil moisture contents increased indicating that dry-down rates were slowing with changing seasons [37]. Additionally, soil dry-down rates were likely at least numerically larger at 10 cm than at 30 cm due to decreased saturated hydraulic conductivities with depth (**Figure 10**, **Table 3**).

3.3 Vegetation differences

Three vegetative properties were affected by one or more or a combination of treatment factors evaluated (i.e., SMR, mound position, and sample date). Total dry matter (DM) differed ($P = 0.04$) between SMRs across mound position over time (**Table 4**). Total DM was numerically greatest (8489 kg ha^{-1}) at the aquic summit in August 2018 and numerically smallest (1280 kg ha^{-1}) at the aquic inter-mound in May 2018 (**Figure 11**). Total DM production was similar at corresponding mound positions between the aquic and udic soils for every treatment combination excluding the aquic and udic mound summits in June and August 2017, in which the udic mound summit produced more DM than the aquic summit (**Figure 11**). Additionally, the mound summit positions generally produced more biomass than the inter-mound positions in both soil moisture regimes (**Figure 11**). Total DM was at least numerically lowest for each respective mound position-SMR combination in May 2018 than all other sampling dates (**Figure 11**).

Averaged across mound position, total DM varied ($P = 0.03$) among SMR-biomass type combinations over time (**Table 5**). Total DM was numerically greatest (5027 kg ha^{-1}) in the aquic-grass combination in August 2018 and numerically least (814 kg ha^{-1}) in the aquic-grass combination in May 2018 (**Figure 11**). For the aquic SMR, grasses significantly outproduced other species on both end-of-season

Source of variation	Grass DM	Forb DM	Total DM	PG
		P		
Soil moisture regime	0.856	0.369	0.128	0.812
Position (P)	0.027	0.097	<0.001	0.951
Sampling date (SD)	<0.001	0.142	<0.001	0.065
SMR × P	0.837	0.156	0.096	0.275
SMR × SD	0.017	0.550	0.161	0.352
P × SD	0.537	0.878	0.071	0.976
SMR × P × SD	0.077	0.956	0.035	0.502

Table 4.
Summary of the combined effects of soil moisture regime (aquic and udic), position (mound summit or inter-mound), and sampling date (2 June 2017, 17 August 2017, 19 May 2018, and 16 August 2018) on grass dry matter (grass DM), forb dry matter (Forb DM), total dry matter (total DM), and percent grass (PG) in a mounded native tallgrass prairie in the Ozark highlands region of Northwest Arkansas.

samples, whereas no differences in DM occurred on either early season sample (**Figure 11**). For the udic soil, grasses outproduced other species in August 2017, with no differences occurring on June 2017 or during the 2018 season (**Figure 11**). Grasses outproducing other plant species is typical of tallgrass prairie ecosystems, as grasses generally account for most of the biomass production and forbs provide species richness and diversity [3, 39].

Averaged across mound position, grass DM differed (P = 0.02) between SMRs over time (**Table 4**). Grass DM was numerically largest (5027 kg ha^{-1}) among all treatment combinations in the aquic soil moisture regime in August 2018 and smallest (814 kg ha^{-1}) in May 2018 in both soil moisture regimes, which did not differ (**Figure 11**). Grass DM was similar between the aquic and udic soils at each sampling date excluding August 2018, in which the aquic soil produced more DM than the udic soil (**Figure 11**). Averaged across SMR and sample date, grass DM differed (P = 0.03) between mound positions, with the mound summit producing 3216 kg ha^{-1} than the 2331 kg ha^{-1} of grass DM in the inter-mound position.

Previous studies analyzing the effect of soil moisture on biomass production have indicated that soil moisture influences plant biomass production. Total above- and belowground biomass had a significant positive correlation with soil moisture content from 0 to 30 cm below the soil surface across 81 grassland ecosystems in the Loess Plateau, China [40]. Similarly, a correlation study conducted by Wu et al. [41] in the Qinghai-Tibetan Plateau, China, concluded that aboveground biomass significantly increased with increased soil moisture in the 0–10- (R^2 = 0.83) and 10–20-cm (R^2 = 0.79) depth intervals. Briggs and Knapp [5] analyzed the influence of soil moisture on biomass production in burned and unburned treatments at Konza Prairie and concluded that soil moisture did not affect grass or forb net primary productivity in long-term unburned watersheds. Conversely, soil moisture at depths <1 m were determined to significantly increase grass and total NPP at annually burned sites [5]. Although the aquic summit generally contained more water than the udic summit, the udic summit produced more total DM than the aquic summit during the 2017 season (**Figure 11**). Increased biomass production in the udic mound summit may have resulted from greater soil organic matter contents to a depth of 60 cm in the udic summit (**Table 2**). Additionally, the aquic and udic inter-mounds exhibited no difference in total DM production, indicating that soil moisture differences between the aquic and udic inter-mound did not affect the biomass production (**Figure 11**).

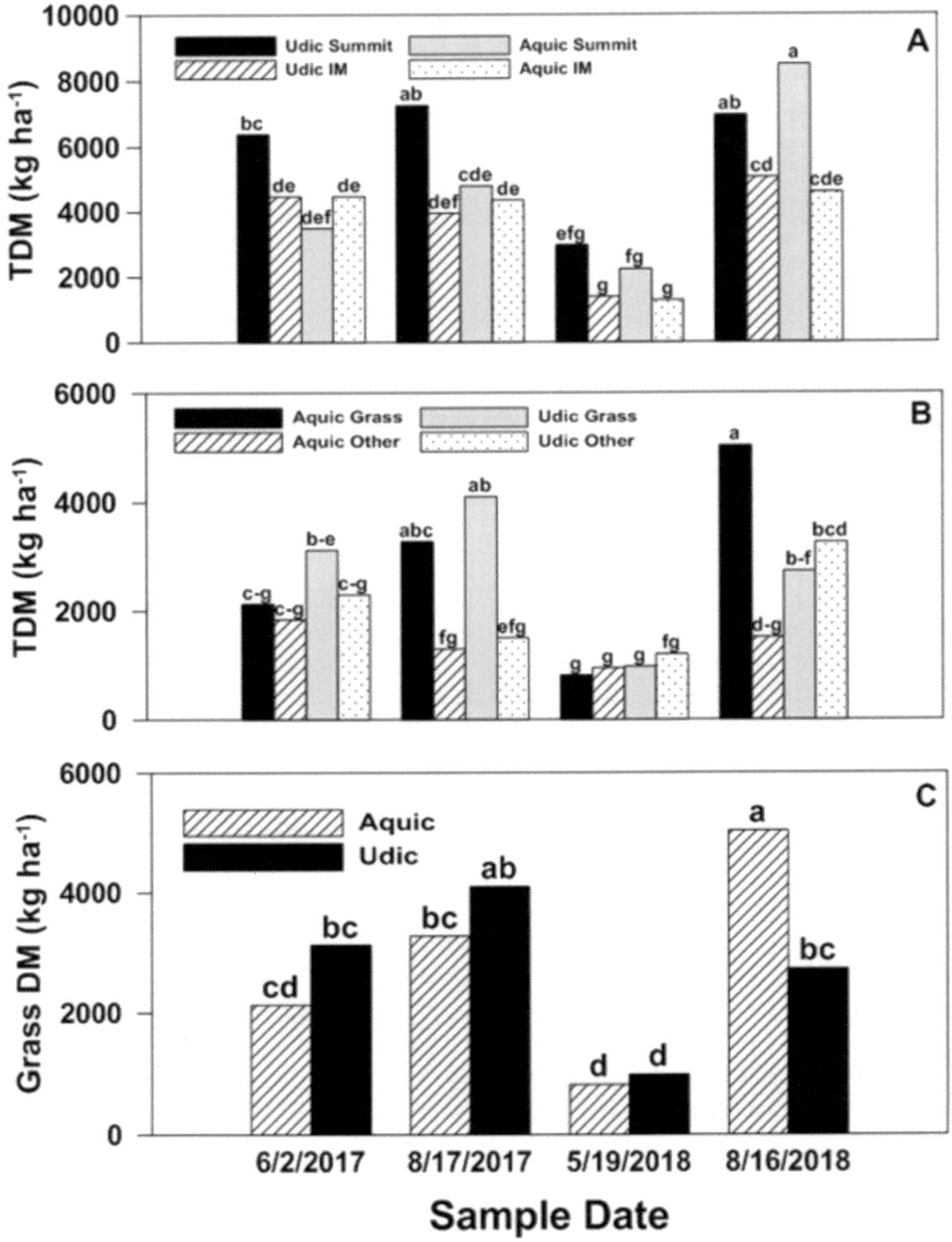

Figure 11.
Sampling date, mound position, and soil moisture regime effects on total dry matter (TDM) (A), sampling date, soil moisture regime, and biomass type effects on total dry matter (TDM) (B), and sampling date and soil moisture regime effects on grass dry matter (grass DM) in a native tallgrass prairie with prairie mounds in the Ozark highlands region of Northwest Arkansas (C).

The difference in DM production between the two early season samples (i.e., June 2017 and May 2018; **Figure 11**) likely resulted from the prescribed burn performed in January 2017. The prescribed burn eliminated dead plant material (i.e., necromass) from the ecosystem and provided the soil with more direct sunlight, which then stimulated plant growth for the 2017 season [42, 43]. The necromass from the increased-biomass-producing 2017 season then shaded the soil surface, slowing soil warming and reducing light availability to newly emerging plants, which would account for the lower total DM production in May 2018 [44].

Source of variation	TDM
	P
Soil moisture regime	0.128
Position (P)	0.011
Sampling date (SD)	<0.001
Biomass type (BT)	0.001
SMR × P	0.377
SMR × SD	0.689
P × SD	0.559
SMR × BT	0.232
P × BT	0.814
SD × BT	0.035
SMR × P × SD	0.457
SMR × P × BT	0.257
SMR × SD × BT	0.028
P × SD × BT	0.969
SMR × P × D × BT	0.341

Table 5.
Summary of the combined effects of soil moisture regime (aquic and udic), position (mound summit and inter-mound), sampling date (2 June 2017, 17 August 2017, 19 May 2018, and 16 August 2018), and biomass type (grass and other species) on total dry matter (TDM) in a mounded native tallgrass prairie in the Ozark Highlands region of northwest Arkansas.

The results of the study are supported by past research analyzing herbage production on mound and inter-mound mound positions [6, 12]. Mounded mound positions seeded with intermediate wheatgrass (*Thinopyrum intermedium*), crested wheatgrass [*Agropyron cristatum* (L.) Gaertn.], smooth brome (*Bromus inermis* Leyss.), Russian wildrye [*Psathyrostachys juncea* (Fisch.) Nevski], and big bluegrass (*Poa secunda* J. Presl) produced 94, 180, 323, 358, and 542% greater herbage yields for the respective plants than the inter-mound positions seeded with the same plants [12]. Additionally, similar to the results of this study, annual forage production on mounded soils in eastern Oklahoma was 4997 kg ha^{-1} compared with 3227 kg ha^{-1} produced by inter-mound soil [10]. Researchers have suggested that mounded soils likely produce larger quantities of biomass than inter-mound soils as a result of enhanced soil fertility and larger quantities of plant-available water present in mounded profiles due to a larger volume of soil in mounded profiles than in inter-mound profiles [12, 45]. According to McGinnies [12], mounded soils contained 66% more nitrogen than inter-mound positions, which would account for mounds producing larger quantities of biomass. At the current site, soil total nitrogen was unaffected by mound position [36]. Soil pH was similar between and across soil moisture regimes for mound summit and inter-mound positions, with exception of the udic mound and udic inter-mound, in which the inter-mound had a more alkaline pH [36]. The water contents in the current study may have been too large in the inter-mounds to promote optimal plant growth, which would explain why herbage production was generally at least numerically larger in the mound summits of both soil moisture regimes. Mound summits had deeper depths to redox features and saturated or near-saturated conditions than the corresponding inter-mound positions, which may have better promoted vegetative growth.

Vegetation comprising the udic mound was 30.7% similar to the vegetation present at the udic inter-mound position. When mound positions were compared across SMRs, the aquic and udic mound summits exhibited 42.8% similarity, whereas the aquic and udic inter-mounds were only 29.6% similar. Lastly, vegetation comprising the aquic mound was 41.3% similar to that of the aquic inter-mound position. The results of the study are supported by previous research analyzing mound summit and inter-mound vegetation [11]. Vegetational similarity between mounds and the adjacent non-mounded prairie area was reported as 35.2% at Kalsow Prairie in Iowa, which is within the range reported in the current study [11]. Scientists have hypothesized the reason for the dissimilarity between the mound positions is due to the microtopographic variation of the mounds compared to inter-mound soils [11, 15]. Del Moral and Deardorff [15] noted that hairy cat's ear (*Hypochaeris radicata* L.) only grew in micro-depressions located on mounds. Additionally, Del Moral and Deardorff [15] determined that plant species, such as racomitrium moss [*Racomitrium canescens* (Hedw.) Brid.], responded to changes in drainage and insolation on mounds, which directly influences soil moisture availability.

Total diversity, species richness, and species evenness were unaffected ($P > 0.05$) by any of the treatment factors (i.e., mound position and SMR) evaluated. Total plant diversity was numerically lowest (0.51) at a udic summit, numerically largest (2.10) at an aquic inter-mound and averaged 1.40 throughout the entire prairie area. Species richness was numerically lowest (5.0) among multiple mound positions, numerically largest (14) at a udic inter-mound, and averaged 7.8 across the entire prairie area. Species evenness was numerically smallest at a udic summit (0.32), numerically largest at an aquic summit (0.89), and averaged 0.70 across the entire prairie area. The plant diversity indices studied may have been influenced by the sampling date. Plants were sampled and identified at the beginning of the growing season, and plant DM during this period was at least numerically lower than on all other sample dates. Additionally, many plants had not yet flowered by this early sampling date. Due to reduced biomass production during the early season, plant diversity indices may have been best represented from plant samples collected during the late-season sample, although the current study still provides valuable insight on plant diversity and species richness and evenness.

Studies analyzing plant species diversity, richness, and evenness in mounded ecosystems are not numerous; however, the results of this study agree with past research comparing plant species richness between mound and inter-mound positions [10, 11, 46]. In eastern Oklahoma, no appreciable difference in species richness occurred between mounded and inter-mound positions, with 18 plants identified in the inter-mound and 13 species identified in the mounded position [10]. Similarly, 51 plant species were identified on mounded positions compared to 49 species in the adjacent prairie at Kalsow Prairie in Iowa [11].

Research has suggested that soil moisture influences plant diversity and species richness and evenness [40, 41]. Across 81 grassland sites in the Loess Plateau of northwestern China, Shannon-Wiener diversity and species richness were significantly and positively correlated with soil water storage in the top 30 cm of soil, while species evenness was correlated to water storage from the 0- to 20-cm depth [40]. Additionally, the Shannon-Wiener diversity, Margalef's index of species richness, and Whittaker's index of species evenness exhibited significant positive relationships with soil water content for seedlings, saplings, and adult tree species in a tropical, dry, deciduous forest in the Vindhyan Highlands, India [47]. Among the various plant growth stages (i.e., seedlings, saplings, and adults), soil water content accounted for 65–77% of the variability in plant diversity, 39–61% of the variability in species richness, and 60–68% of the variability in species evenness [47]. Contrary to the previous studies, despite plant aboveground biomass, vegetative cover, and plant height increasing with soil moisture, plant species richness exhibited an inverse relationship with soil water content in an alpine wetland in the Maqu Wetland Protection Area, China [41]. It was hypothesized that large quantities of soil moisture and species density in the alpine wetlands allowed dominant plant species to outcompete other species, resulting in lower plant species richness [40, 41]. The nonsignificant diversity indices in the current study may be a result of the early season sampling opposed to soil moisture differences, as past research has shown that plant diversity, species richness, and species evenness are directly or inversely related to soil moisture gradients [40, 41, 47].

4. Conclusions

This field study demonstrated that soil hydraulic and vegetative properties differed among soil moisture regime, site position, soil depth, time, and their treatment interactions. The results of this study support the hypothesis that maximum VWCs would increase with depth and be greater in the inter-mound than in the mound positions, whereas VWC minima would be the lowest near the soil surface and in mound positions than being deeper in the profile and in inter-mound positions, respectively. The results of this study support the hypothesis that soil maximum and minimum VWCs would be at least numerically larger in the aquic SMR for a given mound position than in the corresponding udic position. Results did not support the hypotheses that soil dry-down rates would be largest in the mound summits and decrease with depth. However, results partially supported the hypotheses that dry-down rates would be larger in the aquic than udic SMR and be larger during the wet than dry season.

The results of the study did not respectively support the hypotheses that (i) total vegetation diversity would be greatest in the inter-mound and in the udic SMR, (ii) species richness would be greatest at the inter-mound position in both SMRs, and (iii) species evenness would be greatest on the mound summit in both SMRs. Additionally, the hypothesis that total aboveground plant productivity would be greatest on the mound summit than on the inter-mound position on each sampling date and that grasses would be more abundant than other species at mound summit and inter-mound positions at each sampling date was however partially supported. Lastly, results did not support the hypothesis that total aboveground plant productivity would be greater in the aquic than udic SMR at each sampling date.

The study clearly demonstrated that soil volumetric water content and vegetative properties differed among mound positions and between SMRs within the top 50 cm of soil over time. Therefore, prairie management and restoration activities need to account for differing soil moisture regimes and mound topographies in order to be most successful. This study has provided detailed insight into water dynamics and vegetative properties in mounded tallgrass prairie ecosystems; however, additional research detailing soil water contents and vegetation in mounded ecosystems is needed as research on the topic is limited. Research should be continued at Chesney Prairie to monitor the effects of burning the prairie every 3 years on soil physical and chemical properties and vegetation in the mounded tallgrass prairie. Additionally, future research should be focused on identifying additional mounded, native tallgrass prairie fragments to sample across the United States to determine how physical and chemical properties of soil and vegetation in mounded ecosystems differ geographically.

Acknowledgements

We would like to thank Ryder Anderson, Diego Della Lunga, Niyi Omidire, Joe Woolbright, Nate Weston, Theo Witsell, and Matt Sena for their assistance throughout the project.

Conflict of interest

There are no conflicts of interest to declare.

Author details

Tyler J. Durre, Kristofor R. Brye*, Lisa S. Wood and Edward E. Gbur
University of Arkansas, Fayetteville, AR, USA

*Address all correspondence to: kbrye@uark.edu

References

[1] Brye KR, West C, Gbur E. Soil quality differences under native tallgrass prairie across a climosequence in Arkansas. The American Midland Naturalist. 2004;**152**:214-230

[2] Samson F, Knopf F. Prairie conservation in North America. Bioscience. 1994;**44**:418-421

[3] Samson FB, Knopf FL. Prairie Conservation: Preserving North America's most Endangered Ecosystem. Washington, DC: Island Press; 1996

[4] Hinten MT. Combining historical records and geospatial techniques to analyze land use modification in a tallgrass prairie landscape [thesis]. Norman: University of Oklahoma; 2012

[5] Briggs JM, Knapp AK. Interannual variability in primary production in tallgrass prairie: Climate, soil moisture, topographic position, and fire as determinants of aboveground biomass. American Journal of Botany. 1995;**82**:1024-1030

[6] Allgood F, Gray F. Genesis, morphology, and classification of mounded soils in Eastern Oklahoma. Soil Science Society of America Proceedings. 1973;**37**:746-753

[7] Carty DJ, Dixon JB, Wilding LP, Turner FT. Characterization of a pimple mound-intermound soil complex in the Gulf coast prairie region of Texas. Soil Science Society of America Journal. 1988;**52**:1715-1721

[8] Ross B, Tester J, Breckenridge W. Ecology of mima-type mounds in Northwestern Minnesota. Ecology. 1968;**49**:172-177

[9] Lee B, Carter B. Soil morphological characteristics of prairie mounds in the forested region of South-Central United States. In: Proceedings of the 19th World Congress of Soil Science; Brisbane, Australia. Symposium 1.6.1. Brisbane, Australia: IUSS; 2010. pp. 638-641

[10] Allgood F, Gray F. An ecological interpretation for the small mounds in landscapes of Eastern Oklahoma. Journal of Environmental Quality. 1974;**3**:37-41

[11] Brotherson JD. Vegetation of the mima mounds of Kalsow prairie, Iowa. Great Basin Naturalist. 1982;**42**:246-261

[12] McGinnies WJ. Effect of mima-type microrelief on herbage production of five seeded grasses in western Colorado. Journal of Range Management. 1960;**13**:231-234

[13] Mielke HW. Mound building by pocket gophers (Geomyidae): Their impact on soils and vegetation in North America. Journal of Biogeography. 1977;**4**:171-180

[14] Brotherson JD. Species composition, distribution, and phytosociology of Kalsow prairie, a Mesic tall-grass prairie in Iowa. Great Basin Naturalist. 1983;**43**:137-167

[15] Del Moral R, Deardorff DC. Vegetation of the mima mounds, Washington state. Ecology. 1976;**57**:520-530

[16] Brye KR, West CP. Grassland management effects on soil quality in the Ozark highlands. Soil Science. 2005;**170**:63-73

[17] Brye KR, Riley TL, Gbur EE. Prairie restoration effects on soil properties in the Ozark highlands. Electronic Journal of Integrative Biosciences. 2008;**6**:87-104

[18] United States Department of Agriculture, Natural Resource Conservation Service (USDA-NRCS).

MLRA #6. 2017. Available from: https://www.nrcs.usda.gov/wps/portal/nrcs/detail/wv/soils/?cid=nrcs144p2_074201 [Accessed: 04 August 2018]

[19] United States Department of Agriculture, Soil Conservation Service (USDA-SCS). Land Resource Regions and Major Land Resource Areas of the United States. 2006. Available from: https://babel.hathitrust.org/cgi/pt?id=uva.x002265183;view=1up;seq=1 [Accessed: 28 November 2018]

[20] Brye KR, Mersiovsky EP, Hernandez LA, Ward LB. Soils of Arkansas. Fayetteville, AR: University of Arkansas, Division of Agriculture; 2013

[21] Drohan P, Farnham T. 2006. Protecting life's foundation. Soil Science Society of America Journal. 2006;**70**:2086-2096

[22] Cox GW. The distribution and origin of mima mound grasslands in San Diego County, California. Ecology. 1984;**65**:1397-1405

[23] Dalquest WW, Scheffer VB. The origin of the mima mounds of Western Washington. Journal of Geology. 1942;**50**:68-84

[24] Guccione MJ, Shingleur RA, Van Arsdale RB, Pratt M. Origin of some prairie mounds in northwestern Arkansas. In: Abstracts with Programs Geological Society of America, Boulder, CO. 1991;**23**:28

[25] Quinn JH. Prairie mounds of Arkansas. Arkansas Archaeological Society. 1961;**2**:1-8

[26] Archuleta TE. Analysis of mima mounds in Northeastern Arkansas [thesis]. New Orleans: University of New Orleans; 1980

[27] Seifert CL, Cox RT, Forman SL, Foti TL, Wasklewicz TA, McColgan AT. Relict nebkhas (pimple mounds) record prolonged late Holocene drought in the forested region of South-Central United States. Quaternary Research. 2009;**71**:329-339

[28] Arkansas National Heritage Commission (ANHC). Chesney Prairie Natural Area. 2018. Available from: http://www.naturalheritage.com/natural-areas/chesney-prairie-natural-area [Accessed: 08 August 2018]

[29] Holiman B, Woolbright J, James D. Increasing Grassland Block Size and Restoring Ozark Prairie and Woodlands. Arkansas State Wildlife Grant Pre-proposal. 2013. Available from http://www.wildlifearkansas.com/proposals/2013PreProposals/ [Accessed: 08 August 2018]

[30] Neal J, Mlodinow M. Birding Chesney Prairie Natural Area. NWA Audubon. 2012. Available from: http://nwaaudubon.tripod.com/webonmediacontents/BIRDING-CHESNEY-PRAIRIE-NATURAL-AREA.pdf [Accessed: 31 August 2018]

[31] Soil Survey Staff (SSS), Natural Resources Conservation Service (NRCS), United States Department of Agriculture (USDA). Jay Series. 2000. Available from: https://soilseries.sc.egov.usda.gov/OSD_Docs/J/JAY.html [Accessed: 07 August 2018]

[32] Soil Survey Staff (SSS), Natural Resources Conservation Service (NRCS), United States Department of Agriculture (USDA). Taloka Series. 2015. Available from: https://soilseries.sc.egov.usda.gov/OSD_Docs/T/TALOKA.html [Accessed: 07 August 2018]

[33] National Centers for Environmental Information (NCEI). Data Tools: 1981-2010 Normals. 1981-2010 Normals|Data Tools | Climate Data Online (CDO) | National Climatic Data Center (NCDC). 2017. Available from: https://www.ncdc.noaa.gov/cdoweb/datatools/normals [Accessed: 25 November 2018]

[34] Spellerberg IF, Fedor PJ. A tribute to Claude Shannon (1961-2001) and a plea for more rigorous use of species richness, species diversity and the 'Shannon-wiener' index. Global Ecology and Biogeography. 2003;**12**:177-179

[35] Illston BG, Basara JB, Crawford KC. Seasonal to interannual variations of soil moisture measured in Oklahoma. International Journal of Climatology. 2004;**24**:1883-1896

[36] Durre TJ, Brye KR, Wood LS, Gbur EE. Soil moisture regime and mound position effects on soil profile properties in a native tallgrass prairie in northwest Arkansas, USA. Geoderma; 2019;**352**:49-60

[37] Henninger DL, Peterson GW, Engman ET. Surface soil moisture within a watershed—Variations, factors influencing, and relationship to surface runoff. Soil Science Society of America Journal. 1976;**40**:773-776

[38] Scott HD, Wood LS, Miley WM. Long-Term Effects of Tillage on the Retention and Transport of Soil Water. University of Arkansas, Fayetteville: Arkansas Water Resources Center; 1986. Available from: https://scholarworks.uark.edu/cgi/viewcontent.cgi?referer=https://search.yahoo.com/&httpsredir=1&article=1082&context=awrctr [Accessed: 30 December 2018]

[39] Howe HF. Response of early-and late-flowering plants to fire season in experimental prairies. Ecological Applications. 1994;**4**:121-133

[40] Deng L, Wang K, Li J, Zhao G, Shangguan Z. Effect of soil moisture and atmospheric humidity on both plant productivity and diversity of native grasslands across the loess plateau, China. Ecological Engineering. 2016;**94**:525-531

[41] Wu GL, Ren GH, Wang D, Shi ZH, Warrington D. Above- and below-ground response to soil water change in an alpine wetland ecosystem on the Qinghai-Tibetan plateau, China. Journal of Hydrology. 2013;**476**:120-127

[42] Abrams MD. Effects of burning regime on buried seed banks and canopy coverage in a Kansas tallgrass prairie. The Southwest Naturalist. 1988;**33**:65-70

[43] Brye KR, Norman JM, Gower ST. Assessing the progress of a tallgrass prairie restoration in Southern Wisconsin. The American Midland Naturalist. 2002;**148**:218-235

[44] Hulbert LC. Causes of fire effects in tallgrass prairie. Ecology. 1988;**69**:46-58

[45] Giles LJ. The ecology of the mounds on mima prairie with special reference to Douglas-Fir invasion [thesis]. Seattle: University of Washington; 1970

[46] Murray CL. A vegetation analysis of a pimpled prairie in northeastern Oklahoma [thesis]. Tulsa: University of Tulsa; 1974

[47] Chaturvedi RK, Raghubanshi AS. Effect of soil moisture on composition and diversity of trees in tropical dry forest. MOJ Ecology and Environmental Sciences. 2018;**3**:0059. DOI: 10.15406/mojes.2018.03.00059

[48] Arkansas GIS Office. Search Data. 2018. Available from: http://gis.arkansas.gov/ [Accessed: 12 November 2018]

Chapter 3

Effects of Fire on Grassland Soils and Water: A Review

Daniel George Neary and Jackson McMichael Leonard

Abstract

Grasslands occur on all of the continents. They collectively constitute the largest ecosystem in the world, making up 40.5% of the terrestrial land area, excluding Greenland and Antarctica. Grasslands are not entirely natural because they have formed and developed under natural and anthropogenic pressures. Their importance now is to the variety of ecosystem services that they provide: livestock grazing areas, water catchments, biodiversity reserves, tourism sites, recreation areas, religious sites, wild food sources, and natural medicine sources. An important function of grasslands is their sequestration and storage of carbon (C). Mollisol soils of grasslands have deep organic matter horizons that make this vegetation type almost as important as forests for C fixation and storage. Fire has been and continues to be an important disturbance in grassland evolution and management. Natural wildfires have been a component of grasslands for over 300 million years and were important in creating and maintaining most of these ecosystems. Humans ignited fires over many millennia to improve habitat for animals and livestock. Prescribed fire practiced by humans is a component of modern grassland management. The incidence of wildfires in grasslands continues to grow as an issue as droughts persist in semi-arid regions. Knowledge of fire effects on grasslands has risen in importance to land managers because fire, as a disturbance process, is an integral part of the concept of ecosystem management and restoration ecology. Fire is an intrusive disturbance in both managed and wildland forests and grasslands. It initiates changes in ecosystems that affect the composition, structure, and patterns of vegetation on the landscape. It also affects the soil and water resources of ecosystems that are critical to overall ecosystem functions and processes.

Keywords: wildfire, prescribed fire, fire severity, watershed impacts, grasslands

1. Introduction

Fire is a dynamic ecosystem process, generally predictable but uncertain in its timing and occurrence on landscapes [1]. It is an integral component of most wildland forest ecosystems as well as wild and managed grasslands (**Figure 1**). It has been a factor in shaping plant communities for over 300 million years, as long as vegetation and lightning have existed on earth [2–5]. Both managed and wild grasslands are susceptible to localized and widespread fires if climate conditions (drought and wind) are conducive to fire spread [6, 7]. Wildland fire covers a spectrum from low severity, prescribed and grassland fires, to landscape-level

IntechOpen

Figure 1.
Examples of grassland ecosystems: (a) Great Plains Comanche National Grasslands, Colorado (photo courtesy of the Pike and San Isabel National Forest), (b) Emory oak savanna, Coronado National Forest, Arizona, (photo by Daniel G. Neary, USDA Forest Service), (c) high altitude grassland, Apache-Sitgreaves National Forest, Arizona, (photo by Daniel G. Neary, USDA Forest Service), (d) Australia farm, Victoria (photo by Daniel G. Neary, USDA Forest Service), (e) Valparaiso, Chile, watershed (photo by Daniel G. Neary, USDA Forest Service), (f) riparian grassland, Arizona (photo by Daniel G. Neary, USDA Forest Service).

high severity wildfires that affect vegetation, soils, water, fauna, air, and cultural resources [8–12]. Knowledge of fire effects has risen in importance to land managers because fire, as a disturbance process, is an integral part of the concept of ecosystem management and restoration ecology. Fire is an intrusive disturbance in both managed and wildland forests and grasslands. It initiates changes in ecosystems that affect the composition, structure, and patterns of vegetation on the landscape. It also affects the soil and water resources of ecosystems that are critical to overall ecosystem functions and processes [1, 10].

Recycling of carbon (C) and nutrients depends on biological decomposition and fire. In regions where decay is constrained either by dry or cold climates or saturated conditions, fire plays a dominant role in recycling organic matter [1].

In warmer, moist climates, decay plays the dominant role in organic matter recycling, except in soils that are predominantly saturated. However, fires do cause hydrologic and physical changes in these soils [13]. Fire in grasslands affects mainly the aboveground components of vegetation and normally does little damage to the large underground reservoir of organic matter in Mollisols typically found in grassland ecosystems (**Figure 2**, [14]). These soils have typically 80–90% of their carbon pool below ground, away from most of the damaging effects of grassland fires.

Soil is the earth's layer of mineral and organic matter, unconsolidated at the interface between atmosphere and geosphere. It results from physico-chemical and biological processes operating simultaneously and over a long period of time on original geological material [15]. Soil is formed by continual interaction between the soil system and the biotic (faunal and floral), climatic (atmospheric and hydrologic), and topographic components of the environment [1]. Grassland soils are unique because of the herbaceous fine root turnover that contributes to the development of deep, organic matter rich, "A" horizons (**Figure 3**).

Soil is variously integrated to other ecosystem components. It provides plants with air, water, nutrients and also mechanical support for subsistence [10]. It also receives and filters rainfall. In this way, it somewhat defines the portion that evaporates on the surface and the portion that is stored belowground to be slowly drained from upstream slopes to the channels, as well as the portion retained and used for soil processes (e.g., sweating, leaching, etc.). As soon as the infiltration capacity of soil precipitation increases, organic and inorganic surface particles are eroded and end up as sediments, nutrients and pollutants in watercourses that affect water quality. An uninterrupted active movement of gas also occur within the soil and it atmosphere. Soil also provides a repository for many cultural artifacts, which can remain in the soil for thousands of years without undergoing appreciable change [16].

Fire can produce a wide range of changes in landscape appearance but the degree of change and duration in grasslands is usually much less than in forested ecosystems [1]. Grass recovery is usually so rapid that the occurrence of fire is masked within 1 year by rapid regrowth. The fire-induced changes coupled with burn intensities generate varied responses in the water, soil, flowers and fauna of burned ecosystems due to the co-variation between fire severity and ecosystem resonance. There are instantaneous and sustainable reactions to wildfires. Immediate effects result from the combustion of biomass and the release of chemicals in the ash created by fire. The response of biological elements (soil microorganisms and ecosystem vegetation) to these disruptions is both drastic and accelerated.

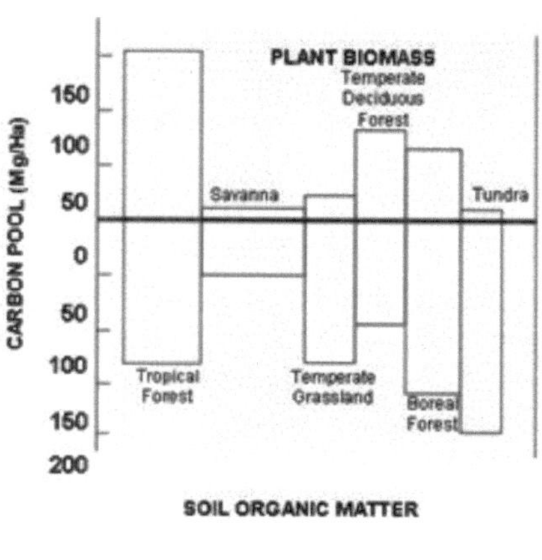

Figure 2.
Distribution of C and soil organic matter (including litter) in major ecosystem types of the world [49].

Figure 3.
Mollisol soil typical of grasslands throughout the world. The notable feature is the organic-rich dark A1 and A2 horizons in the upper 10 to 30 cm formed by grass fine root decomposition and organic matter accumulation in weathered limestone. The deeper white and gold-colored horizon (Cr) is the weathered sedimentary parent material. (photo courtesy of the USDA Natural Resources Conservation Service).

Figure 4.
Wallow fire, Arizona, 2011 (photo courtesy of the USDA Forest Service, Apache-Sitgreaves National Forest).

Fire also releases gas, particles and air pollutants through the combustion of biomass and soil organic matter (**Figure 4**). It thus affects air quality in large airsheds [9, 17]. The long-term fire effects on soils and water are usually subtle, can persist for years following the fire, or can be permanent [10]. Other long-term fire effects arise from the relationships between fire, soils, hydrology, nutrient cycling, and site productivity [18].

Therefore, the objective of this chapter is to examine the effects of fire on natural and managed grassland ecosystems. Fire has been and continues to be an important disturbance in grassland evolution and management. Natural wildfires have been a component of grasslands for millions of years and were important in creating many of these grass and herbaceous ecosystems. Humans ignited fires over many millennia to improve habitat for animals and livestock. Prescribed fire is a component of modern grassland management. The incidence of wildfires in grasslands continues to grow as droughts persist in semi-arid regions of the world [19].

2. Grassland ecosystems

Grasslands collectively constitute the largest ecosystem in the world [20]. They make up 40.5% of the terrestrial land area, excluding Greenland and Antarctica. The areas of grass in the latter two ice-dominated land masses are minor in the global context. Grasslands are divided into woody savannahs and savannahs (13.8%), open and closed shrublands (12.7%), non woody grasslands (8.3%) and tundra (5.7%). The UNESCO definition of grasslands is "lands covered with herbaceous plants that are covered with <10% trees and shrubs". Wooded grasslands typically have 10 to 40% tree and shrub cover [21].

Grasslands are not entirely natural because they have formed and developed under natural (e.g. fire, wild herbivore grazing) and anthropogenic pressures (e.g. prescribed and wildfire, livestock grazing, woody vegetation clearing, over-sowing with pasture grass, etc.) [20]. Grasslands are not considered to be natural if they have been subject to plowing and grass seeding. Their importance now is to the variety of ecosystem services that they provide: livestock grazing areas, water catchments, biodiversity reserves, tourism sites, recreation areas, religious sites, wild food sources, and natural medicine sources. An important function of grasslands is their sequestration and storage of C [22]. Grasslands are characterized by Mollisols, soils with deep organic matter horizons (**Figure 3**) [14]. This vegetation type is almost as important as forests for C fixation and storage. Grassland soils are organic matter sinks on the same order of magnitude as tree biomass.

A question that often arises in grassland management is: Are livestock grazing and burning compatible? Studies in Africa grasslands have demonstrated that fire (natural and prescribed) is often essential in grasslands for maintaining livestock forage in a state critical for herbivores [23]. The negative side of the combined practices is that the amounts of C sequestered in these ecosystems can be reduced.

Fire is typically used as a tool to kill undesirable brush, prevent invasions of poor native species or exotics, and increase forage production [24]. Prescribed burning is used especially on tall-grass prairies of the USA central Great Plains, but timing is important [25]. Cool season grass production can be decreased by spring burns. These early burns can increase summer forage in some species but reduce autumn productivity [26]. Late winter burns have been shown to initiate spring growth 2–3 weeks earlier [27]. Although wildfires have always been a constant part of the prairie fire regimes [6], wildfire numbers and area burned have surged in the 21st Century [28]. Wildfire numbers in the Great Plains of North America increased from 33.4 year^{-1} in 1985 to 116.8 year^{-1} in 2014. The total burned area grew by 400% over the same time period.

In grasslands of the Russian steppe, fire has been used as a tool for millennia to augment natural wildfire starts [29]. Some research has indicated that the treeless steppe is predominantly a result of fire, not soil conditions or climate [3]. Burning is done to remove dead vegetation from the previous year and dry the soil out quicker in the spring. Regrowth after spring fires in the Russian steppes has shown to be richer in nutrient content but lower in yield and cover with decreased soil moisture content [20].

3. Fire characteristics

3.1 Fire regime

The overall fire behavior in a particular type of cover or ecosystem during long successions is defined as the fire mode. This mode describes the typical severity of a fire. But it is recognized that on occasion, relatively severe fires also occur in one type of cover. For example, a stand renewal crown fire is common in forests with large fire-return intervals. The fire regime concept is useful for comparing the relative role of fire between ecosystems and for describing the degree of departure from historical conditions [30, 31]. The fire regime classification used here contains a discussion of the development of fire regime classifications based on fire characteristics and effects [32], combinations of factors including fire frequency, periodicity, intensity, size, pattern, season, and depth of burn, severity, and fire periodicity [10].

The fire modes presented in **Table 1** are described below [10]:

- Understory Fire: These fires are globally not lethal to the dominant canopy and do not significantly change the physiognomy of the predominant vegetation. At least 80% of the dominant vegetation prevails over fires. This mode applies to some types of fire-resistant forests and wooded formations, but also describes the fire regime for true savanna ecosystems (**Figure 5**).

- Stand Replacement Fire: These fires are lethal to most trees, shrub, and grass stems. Greater than 80% of the aboveground dominant vegetation is either incinerated or dies as a result of fire, substantially changing the aboveground vegetative structure. This mode applies to fire-susceptible forests and woodlands, shrublands, and grasslands. In the case of the latter, recovery is very swift within 1 year as grasses sprout from prolific root systems.

- Mixed Fire: The severity of fires varies between nonlethal understory and lethal stand replacement fires. The variation occurs in space or time. First, spatial variability occurs when fire severity varies, producing a spectrum from understory burning to stand replacement within an individual fire. The ultimate result is a fine pattern of vegetation patches. This type of fire regime commonly occurs in some ecosystems because of fluctuations in the fire environment [1, 33]. Complex landscapes facilitate alternating fire severity due to the moisture of the fuel and wind that varies spatially. Temporal change in fire severity results from alternation of individual low intensity, non-repetitive surface fires and long interval stand replacement fires. The result is a variable

Fire Regime Group	Frequency (years)	Severity	Severity and Effects	Fire Regime
I	0-35	Low	Understory Fire	1
II	0-35	Stand Replacement	Stand Replacement	2
III	35-100	Mixed	Mixed	3
IV	35-100+	Stand Replacement	Stand Replacement	2
V	>200	Stand Replacement	Stand Replacement	2
			Non Fire	4

Table 1.
Comparison of fire regime classifications [17, 30, 34].

fire regime [33, 34]. Temporal variability also occurs when cool-moist climate cycles are followed by warm dry ones, leading to multiple decade changes in the role of fire [35].

- Non-fire: In this regime, fire is not likely to occur. However, in rare instances fires do burn in these vegetation types. Although frequently very wet, swamps such as the Everglades in Florida, the Okefenokee Swamp in Georgia, and wet savannas in Brazil have "grassland" areas and have carried wildfires during droughts in the past [36].

3.2 Fire severity

The effects of a specific fire can be described at the stand and community level [1, 33, 37]. The concept describing the ecological and physical effects of a type of fire is the severity. This term indicates the degree of modification and, by extension, the degree of change in ecosystem components. Fire affects both overhead and underground ecosystems components. Most of the effects are above ground level. This is particularly true in grassland fires (**Figure 6**). The degree of fire severity is also related to the vegetation type. For example, in grasslands the differences between prescribed fire and wildfire are normally small due to the lower fuel loads and short duration of burning. In forested environments, the magnitude of the

Figure 5.
Post-fire results in an Emory oak savanna in Southwest New Mexico, 2008. (photo courtesy of A. Kauffman, University of Arizona).

Figure 6.
Wildfire in the Kiowa National Grassland, Cibola National Forest, New Mexico. (photo courtesy of the USDA Forest Service).

effects of fire on water is much lower after a prescribed fire than after a wildfire because of the larger amount of fuel consumed in a wildfire and the heat generated [10]. Canopy-consuming wildfires would be expected to be of the most concern to land managers because of the loss of canopy coupled with the destruction of soil properties. These losses present the worst-case scenario in terms of watershed function impacts.

There is confusion in both the literature and media reports between the terms fire intensity and fire severity. Fire professionals trained in the United States and Canada in fire behavior prediction systems use the term fire intensity in a strict thermodynamic sense to describe the rate of energy released per unit length of fire line [38, 39]. Fire intensity is concerned mainly with the rate of aboveground fuel consumption and energy release rate [37]. The faster a given quantity of fuel burns, the greater the intensity and the shorter the duration [10]. Because the rate at which energy can be transmitted through the soil is limited by the soil's thermal properties, the duration of burning is critically important to the effects on soils [40]. Fire intensity is not necessarily related to the total amount of energy produced during the burning process. Most energy released by flaming combustion of aboveground fuels is not transmitted downward. Only about 5% of the heat from a surface fire is transmitted to the soil [40]. Thus, fire intensity is not always an indicator for measuring the energy transmitted to the soil, nor the physico-chemical and biological changes of the latter. Thus, it is likely that a high intensity, high velocity grass fire would consume little surface litter due to the low energy that is dissipated to the ground during fuel combustion and discharged onto the litter [33]. Only about 5% of the heat released by a surface fire is transmitted into the ground [40]. Therefore, fire intensity is not necessarily a good measure of the amount of energy transmitted downward into the soil, or the associated changes that occur in physical, chemical, and biological properties of the soil. For example, it is possible that a high intensity and fast moving crown fire will consume little of the surface litter because only a small amount of the energy released during the combustion of fuels is transferred downward to the litter surface [33]. In this situation, the resulting surface litter is blackened (charred), but not consumed. Wind driven grassland wildfires will do the same for different reasons.

Because the actual energy released in a fire is not easily measured, the term fire intensity has limited practical application when evaluating ecosystem responses to fire. Fire severity is used more appropriately to indicate the effects of fire on the different ecosystem components [10, 32, 33]. Fire severity has been used describe the magnitude of negative fire impacts on natural ecosystems [41]. It can be seen in a post-fire environment in contrast to intensity which can only be estimated indirectly (**Figure 7**). A further extension of the concept to cover all fire phenomena has been suggested [1, 10]. In this case, severity is a measure of the disruption or damage extent to resources caused by a fire and does not always mean that the phenomenon has negative connotations. Thus, a less severe fire can restore and maintain a range of ecological characteristics that are generally perceived as positive, such as a burning in a pasture. In contrast, a high severity fire in a forest may be a dominant disturbance. In a non-fire adapted ecosystem, it becomes an abnormal, destructive event with long-term consequences.

3.3 Depth of burn

The relationship of fire intensity to fire severity remains largely undefined because of difficulties encountered in relating resource responses to the burning process [33]. While quantitative relationships have been developed to describe changes in the thermal conductivity of soil, and changes in soil temperature and

Figure 7.
Fire severity indicated by color differences (orange spots high severity and black or gray spots moderate severity) and consumption of vegetation, Brins Fire, Arizona. (photo by Daniel G. Neary, Rocky Mountain Research Station, USDA Forest Service).

water content beneath surface and ground fires, these relationships have not been thoroughly extrapolated to field conditions [40, 42]. It is not always possible to estimate the effects of fire on soil, vegetation, and air when these effects are judged by only fire intensity measurements because other factors overwhelm fire behavior. For instance, grassland fires are often high intensity, but their residence time at any one point is extremely short (<15 seconds). The effects on soils are minimal but they are often fatal to animals and humans [43].

The range of fire effects on soil resources can be expected to vary directly with the depth of burn as reflected in the amount of surface litter, organic soil horizons, and woody fuel consumed [10, 33]. Thus, for example, the depth of lethal heat (approximately 60°C) penetration into the soil can be expected to increase with the increasing depth of surface organic material that is burned. In grassland soils this depth is usually very minimal.

The depth of burn into the organic soil horizons, visual observation of charring, and combustion of plant materials defines fire severity for interpreting the effects of fire on soils, plants, and early succession [1, 10, 37] Depth of burn is directly related to the length of time of burning and fuel load in woody fuels [44] and litter [45]. Burn depth can be classified on the basis of visual observation of the degree of fuel consumption and charring on residual plant and soil surfaces [33, 46].

A 1985 summary on the relationships between depth of burn and the charring of plant materials has been updated into a table to reflect subsequent literature [1, 37]. **Table 2** can be used as a guide to classifying depth of burn. The characteristic classes are provided for clarification of subsequent discussion of fire effects:

- Unburned: Plant parts are green and unaltered, there is no direct effect from heat.

- Scorched: Fire did not burn the area, but radiated or convected heat from adjacent burned areas caused visible damage. Soil heating is negligible. Scorched areas occur to varying degrees along the edges of more severely burned areas.

- Light: This class is mostly applicable grasslands but can be found in shrubland and forest types. Plants are charred or consumed, but herbaceous plant bases are not deeply burned, and are still identifiable. Charring of the mineral soil is negligible. Light depth of burn is associated with short duration fires either

Burn class	Category	Description
Unburned	**Surface**	Fire did not burn on the surface.
	Fuels	Some vegetation injury may occur from radiated or convected heat resulting in an increase in dead fuel mass.
	Occurrence	A wide range exists in the percent unburned in natural fuels. Under marginal surface fire conditions the area may be >50%. Under severe burning conditions <5% is unburned. 10 to 20% of the area in slash burns is unburned.
Light	**Surface**	Leaf litter is charred or consumed but some plant parts are discernable. Herbaceous stubble extends above the soil surface. Some plant parts may still be standing, bases not deeply burned, and still recognizable. Surface is black after fire. Charring is limited to <0.2 cm into the soil.
	Fuels	Typically, 50 to 90% of herbaceous fuels are consumed and much of the remaining fuel is charred.
	Occurrence	Burns are spotty to uniform, depending on grass continuity. Light depth of burn occurs in grasslands when soil moisture is high, fuels are sparse, or fires burn under high wind. This is the dominant type of grassland burning.
Moderate	**Surface**	In upland grasslands litter is consumed. Charring extends to <0.5 cm into mineral soil, otherwise soil not altered. Gray or white ash quickly disappears. In grasslands, sedge meadows and prairies growing on organic soils, moderate fires partially burn the root-mat.
	Fuels	Herbaceous plants are consumed to the ground-line.
	Occurrence	Moderate depth of burn tends to occur when soil moisture is low and fuels are continuous. Then burns tend to be uniform. In discontinuous fuels, high winds are required for high coverage in moderate depth of burn.
Deep	**Surface**	In grasslands growing on mineral soil the litter is completely consumed leaving a fluffy white ash surface that soon disappears. Charring to depth of 1 cm in mineral soil. Soil structure slightly altered. In grasslands growing on deep organic soils, fires burn the root-mat and the underlying peat or muck to varying depths.
	Fuels	All above ground fuel is consumed to charcoal and ash.
	Occurrence	In uplands, deep depth of burn is limited to areas beneath the occasional log or anthropogenic features (e.g., fences, corrals). In wetland grasslands, deep burns can occur over large areas when the water table is drawn down.

Table 2.
Depth of burn classes for grasslands [10].

because of light fuel loads, high winds and fire spread, wet fuels, or a combination of these three factors.

- Moderate: This class is mostly applicable to forests but can occur in grasslands or shrublands that are savannas. In grasslands, plants are consumed, and herbaceous plant bases are deeply burned and unidentifiable. In shrublands or woodlands with grass components, average char-depth of the mineral soil is on the order of less than 1 cm, but soil texture and structure are not noticeably altered. Charring may extend up to 3.0 cm beneath shrubs with deep leaf litter in savanna grasslands (**Figure 5**).

- Deep: This class of depth of burn is limited to forests since fuel loads and fire residence times in grasslands are insufficient to produce deep burns. The only exception would be wet savanna grasslands and swamps.

Surface litter and organic soil horizon consumption is a complex process [45]. Depth, bulk density, fire load, mineral filler, moisture rate and wind velocity all affect the heat emission and understory heating. Since these factors are not easily accessed after a fire, it is not possible to define the post-fire criteria potential for discriminating the class. Although analyzing soil calcination is not sufficient to classify the burning severity, the actual depth can be inferred from the preponderance of findings, including the re-composition of the pre-fire plant stratum. Analyses of soil properties, combustion and charring depth of residual vegetation are potential tools to classify the burning extent from the traits presented in **Table 2**.

4. Soil effects

4.1 Soil heating

The energy generated during the ignition and burning of fuels is the driving force behind the physico-chemical and biological alterations in a soil under warming [10]. The soil thermal transmission phenomena include radiation, conduction, convection, mass transport, vaporization and condensation.

- Radiation is defined as the transfer of heat from one body to another, not in contact with it, by electromagnetic wave motion. Radiated energy flows outward in all directions from the emitting substance until it encounters a material capable of absorbing it [1, 10].

- Conduction is the transfer of heat by molecular activity from one part of a substance to another part, or between substances in contact, without appreciable movement or displacement of the substance as a whole [1, 10].

- Convection is a process whereby heat is transferred from one point to another by the mixing of one portion of a fluid with another fluid. Heat transfer by convection plays an important role the rate of fire spread through aboveground fuels.

- Vaporization and condensation are important coupled heat transfer mechanisms that facilitate the rapid transfer of heat through dry soils. Vaporization is the process of heating water until it changes phase from a liquid to a gas. Condensation occurs when a gas is changed into a liquid with heat being released during this process.

Current knowledge on the correlation between soil heating and fire type [10] states the following:

- Crown fires are usually large scale, fast-moving, wind-driven, large, and usually uncontrollable by direct attack. They often have a deep flame front. Generally there is little soil heating when a fire front passes rapidly through the tree crowns. However, if there is sufficient fuel from the forest floor to the crowns, fire will consume all the fuel and produce significant soil heating. These fires are typically occurring in forest canopies and dense shrublands.

- Surface fires which spread slowly, are small-scale, sporadic and controllable, and often having a thick flame front. These fires are capable of igniting and combusting a large part of the forest biomass, bushes and grasslands and can substantially heat the mineral soil.

- Wind-driven grass fires spread very quickly, being sometimes large and marked by a restricted flame front. This is the typical fire of pastoral and wildland ecosystems. The fuel biomass of pastures is less abundant than that of brush and forests and, as a result, soil heating is limited to the passage of surface fires.

- Smoldering fires are flame-free, slow moving and unimpressive, but frequently have long burnout times. These fires are not common in grasslands but can occur, particularly in savannas, where woody roots have extended into grass areas.

4.2 Soil physical properties

The general effects of fire on soil physical properties range from very minor to serious [1, 10]. Since grassland fires are often rapidly-moving with the wind, and have much less fuel than that in brush and forest ecosystems, soil heating is significantly lower, and therefore physical damage much less than what occurs during crown, surface, or smoldering fires. Physical effects include alterations in soil physical properties, development of water repellency, and erosion [47, 48]. Comparisons of wildfire and prescribed fire in grassland ecosystems show little differences in soil impacts due to the low fire severities characteristics of both types of fire.

The amount of aboveground and belowground organic matter varies widely between different vegetation types depending upon on the temperature and moisture conditions prevailing in a particular area. In almost all ecosystems throughout the world, greater quantities of C (a measure of organic matter production) are found belowground than aboveground (**Figure 2**). In grasslands, savannas, and tundra-covered areas, much greater quantities of organic C are found in the underground plant parts of herbaceous vegetation ecosystems (90%) than in aboveground components (<10% of the total C) [49]. Because of the large belowground pools of C in grasslands, fires do not significantly affect the role or importance of C in soil physical properties.

Plant litter is a key factor in determining watershed condition. In grassland ecosystems, easily identifiable layers (Oi, Oe, and Oa layers) may not be present and will be different from forests with thin Oi and Oe horizons and a very deep Oa. The Oi layer consists of freshly fallen plant litter. The Oe layer is made up of partially decomposed litter, and the Oa layer consists of well-decomposed organic matter [50]. Mesic grasslands have a complete herbaceous plant cover and well-developed organic soil horizons (60 cm or more), but in semi-arid environments, the soil is sometimes devoid of cover between the plants. Surface organic matter absorbs precipitation, allowing water to infiltrate deep down instead. Reduction of organic material by high severity fire could result in adverse changes in hydrologic conditions in some instances.

Undisturbed forests have the highest saturated hydraulic conductivities (Ksat) because of their deep soils and high porosities [50]. Grassland soils are similar but not quite as porous as woodlands due to the lack of root-caused macroporosity. Woodlands have the lowest since they are less productive ecosystems and their soils are often lithic, shallow, poorly structured, less permeated by roots and soil organisms that develop macroporosity. Thus the relationships of (Ksat) in these wildland soils is forests > grasslands > woodlands. The importance of surface organic horizons in determining the levels of (Ksat) in forest, woodland, and grassland soils cannot be overstated.

The (Ksat) rates for most undisturbed forests range from 143 to 1000 mm hr^1 [50]. The rates for grasslands reported by in the same paper are 8–612 mm hr^{-1}. Heavily grazed grasslands can have significantly lower rates due to compaction

Location	Reference	Burned condition		Soil depth (cm)
		Saturated hydraulic conductivity K_{sat} (mm hr^{-1})		
		Unburned	Burned (Rx/Wf)	
Australia	[51]	15	34	0–40
Slovakia	[52]	612	972	0–05

Table 3.
Effects of fire in grasslands on soil saturated hydraulic conductivity (K_{sat}). Adapted from [50].

from animal traffic. The highest Ksat rates are usually associated with thick Oe or Oa horizons. A number of recent studies reported changes in Ksat that demonstrate clear reductions in conductivity after fires. Fire severity plays a key role in some of these reductions, but other investigators have demonstrated a surprising lack of correlation with severity and even a reverse trend (**Table 3**; [51, 52]).

An interesting trend emerging out of some of the recent Australian research on the impacts of wildfires on soil hydrologic properties is that the soil surface Ksat values can be similar regardless of severity and that natural water repellency may produce Ksat values less than those measured in burned soils [53]. Seasonal effects are occur when summer natural water repellency disintegrates and the Ksat coefficient reinforces the correlation between control and burnt soils during winter. As a result, fire's gravity—water repellency—hydraulic conductivity ratios are obviously more complex.

4.3 Nutrients

Investigations about the impacts of fire frequency on grazing lands and shrubs have had less desirable outcomes. For example, the annual burning of tall grass prairies in the Great Plains of the Central United States resulted in a significant decrease in soil organic nitrogen (N), microbial biomass, N availability and higher C/N ratios in soil organic matter [54]. Similarly, the increase in available N could harm some nutrient-deficient shrub ecosystems, as previously reported in South Africa. On lowland fynbos, a double increase in soil nutrient content through fire threatens the survival of native species developed on these impoverished ecotopes [55].

4.4 Soil Fauna

The influence of soil heating on earthworms is still little known. The indirect effects of a fire are probably more pronounced than the direct heating of earthworm populations in tallgrass pastures and prairies [56]. Thus, fire intensified the activity of earthworms due to differences in productivity between plants before and after the occurrence of fire. In general, grassland soils are full of roots and rhizomes, which, combined with soil moisture, provide an ideal microclimate for earthworms in the upper 10–20 cm of soil depth. Deep soil protects earthworms from the direct consequences of soil heating during fuel burning, except in the case of severe, long-lasting fires under slash and log piles or in smoldering roots and litter. Grassland fires increase exotic earthworm species at the expense of endemic ones [57].

4.5 Roots and reproductive structures

Many plant roots, regeneration structures and seeds are just above the ground or spread deep down. These plant parts include taproots, surface roots, rhizomes, stolons, root crowns and bulbs. Many roots are in the superficial plant litter layers (L, F and H horizons) and are directly affected when these layers are heated or consumed by fire [10]. Roots of grass plants are less susceptible to damage since they are distributed mainly in the mineral soil and at depths where heating is minimal (**Figure 2**).

Plant roots are sensitive to both duration of heating and the magnitude of the temperature reached. Temperatures of 60°C for 1 minute are sufficient to coagulate protein [58]. Lethal temperatures can occur before proteins began to coagulate. The plant material lethal temperature is very dependent on its moisture content. Wetter plants are prone to destruction at very low temperatures and during shorter warming [59]. Plant roots of grassland soils are well insulated by the soil and have a lower risk of being subjected to lethal temperatures during a fire [1]. The two most important factors that insulate roots against soil heating are their depth in the soil and the soil water content. Generally, the deeper the plant roots are located in the soil, the greater will be the survival rate [60]. Grassland soils have deep "A" horizons formed by fine root turnover and thus are less at risk from low severity, high intensity, and short duration herbaceous fuel fires [14]. Low-severity fire destroys only the surface plant litter aboveground plant structures. In contrast, high-severity fires can consume all the surface organic matter and easily heat the mineral soil above the lethal temperature for roots [1, 10]. This situation typically occurs where woody trees and shrubs have invaded grasslands in the absence of fire. It contributes to the mortality of the woody species and survival of herbaceous plants.

Most seeds are stored in the litter and under the foliage. Medium and intensive fires heat the surface deposits enough to eliminate the seeds deposited there. The deadly seed temperature is about 70°C in wet soils and 90°C in dry soils [61]. Fire can destroy seeds, but it also can enhance reproduction by destroying allelopathic substances that inhibit seed production [62]. Or, fire can provide a mineral seedbed required for new grass germination and growth. The heating associated with fires may also stimulate the germination of seeds that lie dormant in the soil for years because of impermeable seed coats. Regeneration of grass ecosystems after burning is mostly dependent on sprouting from deep and undamaged root systems than seed sprouting.

Soil heating, heat transport and the lethality consequences on seeds and roots are more complex in moist than dry soils [10]. Dry soil is a poor thermal vector and, as a result, heat does not reach deeper into the ground, especially whether the flame front is of short duration. This is the situation with grassland fires. They are incredibly hot, but move rapidly over any one point of ground. The surface of dry soil can easily exceed the lethal temperature of living tissue of roots, while ambient daily soil temperatures can prevail in just 2 cm downward in the soil, with little damage occurring to the roots. Therefore, when the roots of grassland plants are in dry soil, they are not likely to be damaged by wildfire unless the residence time of the flaming front is long.

5. Watershed responses

5.1 General effects

The magnitude of the effects of fire on water quantity and quality are primarily driven by fire severity, and ancillary factors such as post-fire cover, slope, water repellency, and rain fall amount and intensity [1, 10]. Fire intensity is rarely a factor. Fire severity is related to the amount of fuel consumed and resource damage, while fire intensity is only a measure of the rate of heat release. The more severe the fire, the greater the amount of fuel consumed, heat released, soil properties affected, and hydrologic condition altered.

High severity fires increase the amount of nutrients mobilized and alter the hydrologic response of catchments. These combinations of factors make sites more susceptible to erosion of soil and release of nutrients into stream and lakes where they could potentially affect water quality. Wildfires usually are more severe than prescribed fires because of controls over burning conditions and fuel loads in the latter case. As a result, they are more likely to produce significant impacts on watershed resources.

As mentioned earlier, the degree of fire severity is related to the vegetation type. For example, in grasslands the differences between prescribed fire and wildfire are usually small. In forested environments, the magnitude of the effects of fire on water yield and water quality are much lower after a prescribed fire than after a wildfire because of the larger amount of fuel consumed in a wildfire, the greater heat release, and the generally higher severity. Canopy-consuming wildfires are the greatest concern to watershed managers because of the loss of canopy coupled with soil property damage and alterations to hydrologic conditions. The differences between wildfire and prescribed fire in shrublands are intermediate between those seen in grass and forest environments.

In grassland ecosystems, high-severity fires have been shown to increase the amount of nutrients mobilized and alter the hydrologic response of catchments [63]. The combination of these factors makes sites more susceptible to erosion of soil and the release of nutrients into stream and lakes where they could potentially affect water quality. Typically, prescribed fires are less severe than wildfires because of controls over burning conditions and fuel loads. Prescribed fire at Konza Prairie Biological Station, one of the last remaining areas of unplowed, tall-grass prairie in the Midwestern United States, released Dissolved Black Carbon (DBC) into nearby grassland stream systems [64]. The study concluded that there was not a direct relationship between water quality and DBC generated by prescribed burning. They suggest one reason for this is that the export of DBC through stable grassland systems can be on the scale of decades to centuries.

5.2 Scale

Another important factor in the impact on wildfire on watershed function is the size of the fire. Wildfire spread in grasslands is a function of fuel type, fuel moisture, air temperatures, and wind speeds. Influenced by the spotting of embers out ahead of the main fire front, wind-driven fire events in grasslands can move rapidly. Fire spread was modeled in Australian grasslands based on a critical wind speed based on a critical wind speed of 5 km hr^{-1} [65]. A linear relationship was used for rates of spread below wind speeds of 5 km hr^{-1}. Above that speed a power function with an exponent of less than 1 was needed for the model to match field data. Data from 21 grassland fires with wind speeds ranging from 27 to 78 km hr^{-1} produced spread rates of 4 to over 23 km hr^{-1}.

After months of drought, low-humidity and above average temperatures in 2011 a series of wind-driven grassland fires in the state of Texas, U.S.A. consumed nearly 1,618,750 ha during a single fire season, nearly double the previous record [66]. Multiple individual fires were in excess of 50,000 ha in size. Due to their size, fires of this magnitude compound the impacts of large scale erosion events [63]. This sort of situation has the potential to result in large scale degradation of grassland soil nutrients and hydrologic function and lead to desertification on the scale seen in the Great Plains of North America in the 1930s.

A prescribed fire in a Texas grassland resulted in a large increase (1150%) in streamflow in comparison to an unburned watershed in the first year after burning [68]. The increased post-fire streamflow was short lived, however, with flows returning to pre-fire levels shortly after the burning. By contrast, post-fire streamflow increases in forests and shrublands remained elevated for significant number of years because of the delay in revegetation. Grassland vegetation growth after fire is usually quite rapid (a few months to a year).

Another important determinant of the magnitude of the effects of fire on water is slope. Steepness of the slope has a significant influence on movement of soil and nutrients into stream channels where they can affect water quality. A study of the impacts of slope on grassland fires demonstrated that as slope increased in a prescribed fire, erosion from slopes accelerated [67, 68].

Annual streamflow totals (annual water yields) generally increase as precipitation inputs to a watershed increase [10]. Streamflows originating on forest watersheds, therefore, are generally greater than those originating on grassland watersheds, and those from grasslands are greater than flows originating on desert watersheds. Furthermore, annual streamflow totals frequently increase when mature forests are harvested or otherwise cut, attacked by insects, or burned [68, 69]. The observed increases in streamflow following disturbances often diminish with decreasing precipitation inputs to a watershed. This decrease can occur within a year or take many years, depending on the disturbance and vegetation type.

6. Summary and conclusions

Grasslands collectively are the largest ecosystem in the world, making up 40.5% of the land mass excluding Greenland and Antarctica. They are not entirely natural because they have formed and developed under the influence of natural and anthropogenic disturbances fire, (e.g. prescribed fire, wildfire, livestock grazing, woody vegetation clearing, over-sowing with pasture grass, etc.). Their importance now is in the variety of ecosystem services that they provide. A critical function of grasslands in global C circulation is their subsoil sequestration and storage of organic matter. Grasslands soils are classified as Mollisols, soils with deep, organic matter horizons. This characteristic makes grasslands almost as important as forests for C fixation and storage. Grassland soils are organic matter sinks on the same order of magnitude as tree biomass.

Although wildfires have always been a constant part of prairie fire regimes, wildfire numbers and area burned have surged in the 21st Century due to drought. The number of wildfires in the Great Plains of North America increased from 33.4 year^{-1} in 1985 to 116.8 year^{-1} in 2014. The total burned area grew by 400% over the same time period. Measured wind speeds ranging from 27 to 78 km hr^{-1} have produced spread rates of 4 to over 23 km hr^{-1}.

The general effects of fire on soil physical properties range from very minor to serious. Since grassland fires are often rapidly-moving with the wind, and have much less fuel than that in brush and forest ecosystems, soil heating is significantly lower, and therefore physical damage is much less than what occurs during crown, surface, or smoldering fires in forests and woodlands. Physical effects of fire include alterations in soil physical properties, development of water repellency, and erosion. Comparisons of wildfire and prescribed fire in grassland ecosystems show little differences in physical impacts due to the low fire severities characteristics of both types of fire, the narrow flame fronts, and rapid spread rates.

In grasslands, savannas, and tundras, much greater quantities of organic C are found in the underground (90%) than in aboveground components (<10% of the total C). Because of these large belowground pools of C in grasslands, fires do not significantly affect the role or importance of organic C in maintaining the physical and hydrological properties of Mollisol soils.

Acknowledgements

The Authors would like to thank the Rocky Mountain Research Station, Air-Water-Aquatic Environments Research Program, and the Program Manager, Frank McCormick, for support of this effort.

Conflict of interest

There are no "Conflicts of Interest" associated with this paper. It was produced by U.S. Forest Service employees during normal work hours and on appropriated funding.

Author details

Daniel George Neary* and Jackson McMichael Leonard
USDA Forest Service, Rocky Mountain Research Station, Air, Water, Aquatic Environments Program, Flagstaff, Arizona, USA

*Address all correspondence to: dan.neary@usda.gov

References

[1] DeBano LF, Neary DG, Ffolliott PF. Fire's Effects on Ecosystems. New York: John Wiley & Sons, Inc.; 1998. p. 333

[2] Pyne SJ. Fire in America: A Cultural History of Wildland and Rural Fire. Seattle: University of Washington Press; 1982. p. 654

[3] Pyne SJ, Andrews PL, Laven RD. Introduction to Wildland Fire. New York: John Wiley & Sons; 1996. p. 769

[4] Scott AC. The pre-quaternary history of fire. Paleogeography, Paleoclimatology, Paleoecology. 2000;**164**:281-329

[5] Smith JK, Zouhar K, Sutherland S, Brooks ML. Chapter 1: Fire and nonnative invasive plants – Introduction. Pp. 1-31. In: Zouhar K, Smith JK, Sutherland S, Brooks ML, editors. Wildland Fire in Ecosystems: Fire and Nonnative Invasive Plants. General Technical Report RMRS-GTR-42 Volume 6. Ogden, UT: U.S. Department of Agriculture, Forest Service, Rocky Mountain Research Station; 2008. p. 355

[6] Jackson AS. Wildfires in the Great Plains grasslands. In: Komarek EV, editor. Proceedings of the 4th Tall Timbers Fire Ecology Conference. 1965. pp. 241-259

[7] Noble JC. Behaviour of a very fast grassland wildfire on the riverine plain of southeastern Australia. International Journal of Wildland Fire. 2008;**1**:189-196

[8] Smith JK, editor. Wildland Fire in Ecosystems: Effects of Fire on Fauna. General Technical Report RMRS-GTR-42 Volume 1. Ogden, UT: U.S. Department of Agriculture, Forest Service, Rocky Mountain Research Station; 2000. p. 83

[9] Sandberg DV, Ottmar RD, Peterson JL. Wildland Fire in Ecosystems: Effects of Fire on Air. General Technical Report RMRS-GTR-42 Volume 5. Ogden, UT: U.S. Department of Agriculture, Forest Service, Rocky Mountain Research Station; 2002. p. 79

[10] Neary DG, Ryan KC, DeBano LF, editors. Wildland Fire in Ecosystems: Effects of Fire on Soil and Water. General Technical Report RMRS-GTR-42 Volume 4. Ogden, UT: U.S. Department of Agriculture, Forest Service, Rocky Mountain Research Station; 2005. p. 250

[11] Zouhar K, Smith JK, Sutherland S, Brooks ML, editors. Wildland Fire in Ecosystems: Fire and Nonnative Invasive Plants. General Technical Report RMRS-GTR-42 Volume 6. Ogden, UT: U.S. Department of Agriculture, Forest Service, Rocky Mountain Research Station; 2008. p. 355

[12] Bento-Gonçalves A, Vieira A, Úbeda X, Martin D. Fire and soils: Key concepts and recent advances. Geoderma. 2012;**191**:3-13

[13] Watts AC, Schmidt CA, McLaughlin DL, Kaplan DA. Hydrologic implications of smoldering fires in wetland landscapes. Freshwater Science. 2015;**34**:1394-1405

[14] Fenton TE. Chapter 4: Mollisols. In: Developments in Soil Science, Part B. Vol. 11. Amsterdam, Netherlands: Elsevier; 1983. pp. 125-163

[15] Singer MJ, Munns DN. Soils: An Introduction. 3rd ed. Upper Saddle River, NJ: Prentice Hall; 1996. p. 480

[16] Ryan KC, Jones AT, Koerner CL, Lee KM. (Editors) Wildland Fire in Ecosystems: Effects of Fire on Cultural Resources and Archeology.

General Technical Report RMRS-GTR-42 Volume 3. U.S. Department of Agriculture, Forest Service, Rocky Mountain Research Station: Ogden, UT; 2012. p. 224

[17] Hardy CC, Menakis JP, Long DG, Brown JK. Mapping historic fire regimes for the Western United States: Integrating remote sensing and biophysical data. In: Greer JD, editor. Proceedings of the 7th Forest Service Remote Sensing Applications Conference; 1998 April 6-10; Nassau Bay, TX. Bethesda, MD: American Society for Photogrammetry and Remote Sensing; 1998. pp. 288-300

[18] Neary DG, Klopatek CC, DeBano LF, Ffolliott PF. Effects of fire on belowground sustainability: A review and synthesis. Forest Ecology and Management. 1999;**122**:51-71

[19] Brown KJ, Clark JS, Grimm EC, Donovan JJ, Mueller PG, Hansen BCS, et al. Fire cycles in north American interior grasslands and their relation to prairie drought. Proceedings of the National Academy of Sciences. 2005;**102**:8865-8870

[20] Suttie JM, Reynolds SG, Batelio C, editors. Grasslands of the World. Food and Agriculture Organization of the United Nations, Plant Production and Protection Series, No. 342005. p. 514

[21] White F. The Vegetation of Africa: A Descriptive Memoir to Accompany the UNESCO/AETFAT/UNSO Vegetation Map of Africa. Natural Resources Research Series XX. UNESCO: Paris, France; 1983. p. 356

[22] Minahi K, Goudriaan J, Lantinga EA, Kimura T. Significance of grasslands in emission and absorption of greenhouse gases. In: Barker MG, editor. Grasslands for our World. Wellington, New Zealand: SIR Publishing; 1993

[23] Dublin HT. Vegetation dynamics in the Serengeti-Mara ecosystem: The role of elephants, fire, and other factors. In: Sinclair ARE, Arcese P, editors. Serengeti II: Dynamics, Management, and Conservation of an Ecosystem. Chicago, Illinois: University of Chicago Press; 1995. pp. 71-90

[24] Wright HA, Bailey AW. Fire Ecology United States and Southern Canada. New York: John Wiley & Sons; 1982. p. 501

[25] Wright HA. Range burning. Journal of Range Management. 1974;**27**:5-11

[26] Vallentine FF. Range Development and Improvements. 3rd ed. New York: Academic Press; 1989. p. 524

[27] Ehrenreich JH, Aikman JM. An ecological study of certain management practices on native plants in Iowa. Ecological Monographs. 1963;**33**:113-130

[28] Donovan VM, Wonkka C, Twidwell D. Surging wildfire activity in a grassland biome. Geophysical Research Letters. 2017;**44**:5986-5993

[29] Boonman JG, Mikhalev SS. Chapter 10. The Russian steppe. Pp. 381-416. In: Suttie JM, Reynolds SG, Batelio C, editors. Grasslands of the World. Food and Agriculture Organization of the United Nations, Plant Production and Protection Series, No. 34. 2005. p. 514

[30] Hardy CC, Schmidt KM, Menakis JP, Sampson RN. Spatial data for national fire planning and fuel management. International Journal of Wildland Fire. 2001;**10**:353-372

[31] Schmidt KM, Menakis JP, Hardy CC, Hann WJ, Bunnell DL. Development of Coarse-Scale Spatial Data for Wildland Fire and Fuel Management. General Technical Report RMRS-87. U.S. Department of Agriculture, Forest Service, Rocky Mountain Research Station: Fort Collins, CO; 2002. p. 41

[32] Agee JK. Fire Ecology of Pacific Northwest Forests. Washington, DC: Island Press; 1993. p. 493

[33] Ryan KC. Dynamic interactions between forest structure and fire behavior in boreal ecosystems. Silva Fennica. 2002;**36**:13-39

[34] Brown JK, Smith JK, editors. Wildland Fire in Ecosystems: Effects of Fire on Flora. General Technical Report RMRS-GTR-42 Volume 2. Ogden, UT: U.S. Department of Agriculture, Forest Service, Rocky Mountain Research Station; 2000. p. 257

[35] Kauffman JB, Steele MD, Cummings D, Jaramillo VJ. Biomass dynamics associated with deforestation, fire, and conversion to cattle pasture in a Mexican tropical dry forest. Forest Ecology and Management. 2003;**176**:1-12

[36] Schmidt IB, Fidelis A, Miranda HS, Ticktin T. How do the wets burn? Fire behavior and intensity in wet grasslands in the Brazilian savanna. Brazilian Journal of Botany. 2017;**40**:167-175

[37] Feller MC. The influence of fire severity, not fire intensity, on understory vegetation biomass in British Columbia. In: Proceedings, 13th conference on fire and Forest meteorology; 1996 October 27-31; Lorne, Australia. International Journal of Wildland Fire. 1998;7:335-348

[38] Stocks BJ, Lynham TJ, Lawson BD, Alexander ME, Van Wagner CE, McAlpine RS, et al. The Canadian forest fire danger rating system: An overview. The Forestry Chronicle. 1989;**65**:250-257

[39] Alexander ME. Calculating and interpreting forest fire intensities. Canadian Journal of Botany. 1982;**60**:349-357

[40] Campbell GS, Jungbauer JD Jr, Bristow KL, Hungerford RD. Soil temperature and water content beneath a surface fire. Soil Science. 1995;**159**:363-374

[41] Simard AJ. Fire severity, changing scales, and how things hang together. International Journal of Wildland Fire. 1991;**1**:23-34

[42] Campbell GS, Jungbauer JD, Bidlake WR, Hungerford RD. Predicting the effect of temperature on soil thermal conductivity. Soil Science. 1994;**158**:307-313

[43] De Ronde C. Wildland fire-related fatalities in South Africa – A 1994 case study and looking back at the year 2002. P. 158. In: Viegas DX, editor. Forest Fire Research & Wildland Fire Safety. Rotterdam, Netherlands: Millpress; 2002. p. 228

[44] Albini FA, Reinhardt ED. Modeling ignition and burning rate of large woody natural fuels. International Journal of Wildland Fire. 1995;**5**:81-91

[45] Johnson EA, Miyanishi K, editors. Forest Fires, Behavior and Ecological Effects. San Francisco, California: Academic Press; 2001. p. 594

[46] Ryan KC, Noste NV. Evaluating prescribed fires. In: Lotan JE, Kilgore BM, Fischer WC, Mutch RW, editors. Proceedings—Symposium and Workshop on Wilderness Fire. General Technical Report INT-182. Ogden, UT: U.S. Department of Agriculture, Forest Service, Intermountain Forest and Range Experiment Station; 1985. pp. 230-238

[47] DeBano LF. The role of fire and soil heating on water repellency in wildland environments: A review. Journal of Hydrology. 2000;**231-232**:195-206

[48] Doerr SH, Shakesby RA, Walsh RPD. Soil water repellency: Its causes, characteristics and hydro-geomorphological significance. Earth Science Reviews. 2000;**51**:33-65

[49] Anderson JM. The effects of climate change on decomposition processes in grassland and coniferous forest. Ecological Applications. 1991;**1**:326-347

[50] Neary DG. Impacts of wildfire severity on hydraulic conductivity in forest, woodland, and grassland soils. Pp. 123-142. In: Elango L, editor. Hydraulic Conductivity – Issues, Determination, and Applications. Rijeka, Croatia: INTECH; 2011. p. 424

[51] Valzano FP, Greens RSB, Murphy BW. Direct effects of stubble burning on soil hydraulic and physical properties in a direct drill tillage system. Soil& Tillage Research. 1997;**42**:209-219

[52] Novák V, Lichner L, Zhang B, Kňava K. The impact of heating on the hydraulic properties of soils sampled under different plant cover. Biologia. 2009;**64**:483-486

[53] Sheridan G, Lane PNJ, Noske PJ. Quantification of hillslope runoff and erosion processes before and after wildfire in a wet eucalyptus forest. Journal of Hydrology. 2007;**343**:12-28

[54] Ojima DS, Schimel DS, Parton WJ, Owensby CE. Long and short-term effects of fire on nitrogen cycling in tallgrass prairie. Biogeochemistry. 1994;**24**:67-84

[55] Musil CF, Midgley GF. The relative impact of invasive Australian acacias, fire and season on the soil chemical status of a sand plain lowland fynbos community. South African Journal of Botany. 1990;**56**:419-427

[56] James SW. Effects of fire and soil type on earthworm populations in a tallgrass prairie. Pedobiologia. 1982;**24**:140-147

[57] Callaham MA Jr, Hendrix PF, Phillips RJ. Occurrence of an exotic earthworm (Amynthas agrestis) in undisturbed soils of the southern Appalachian Mountains, USA. Pedobiologia. 2003;**47**:466-470

[58] Precht J, Chrisphersen J, Hensel H, Larcher W. Temperature and Life. New York: Springer-Verlag; 1973. p. 779

[59] Zwolinski MJ. Fire effects on vegetation and succession. In: Krammes MJ, editor. Proceedings of a Symposium Effects of Fire Management of Southwestern Natural Resources. General Technical Report RM-191. Fort Collins, CO: U.S. Department of Agriculture, Forest Service, Rocky Mountain Forest and Range Experiment Station; 1990. pp. 18-24

[60] Flinn MA, Wein RW. Depth of underground plant organs and theoretical survival during fire. Canadian Journal of Botany. 1977;**55**:2550-2554

[61] Martin RE, Miller RL, Cushwa CT. Germination response of legume seeds subjected to moist and dry heat. Ecology. 1975;**56**:1441-1445

[62] Miller M. Chapter 2: Fire autecology. Pp 9-34. In: Brown JK, Smith JK, editors. Wildland Fire in Ecosystems: Effects of Fire on Flora. General Technical Report RMRS-GTR-42 Volume 2. Ogden, UT: U.S. Department of Agriculture, Forest Service, Rocky Mountain Research Station; 2000. p. 257

[63] Ravi S, D'Odorico P, Zobeck TM, Over TM. The effect of fire-induced soil hydrophobicity on wind erosion in a semiarid grassland: Experimental observations and theoretical framework. Geomorphology. 2009;**105**:80-86

[64] Ding Y, Yamashita Y, Dodds WK, Jaffé R. Dissolved black carbon in grassland streams: Is there an effect of recent fire history? Chemosphere. 2013;**90**:2557-2562

[65] Cheney NP, Gould JS, Catchpole WR. Prediction of fire spread in grasslands. International Journal of Wildland Fire. 1998;**8**:1-13

[66] Texas Forest Service; 2011. Available from: https://ticc.tamu.edu/Documents/Home/tx_sitrep.pdf [Accessed: 24 March 2019]

[67] Wright HA, Churchill FM, Stevens WC. Effect of prescribed burning on sediment, water yield, and water quality from juniper lands in Central Texas. Journal of Range Management. 1976;**29**:294-298

[68] Bosch JM, Hewlett JD. A review of catchment experiments to determine the effect of vegetation changes on water yield and evapotranspiration. Journal of Hydrology. 1982;**55**:3-23

[69] Hornbeck JW, Adams MB, Corbett ES, Verry ES, Lynch JA. Long-term impacts of forest treatments on water yields: A summary for northeastern USA. Journal of Hydrology. 1993;**150**:323-344

Chapter 4

Increasing Warm-Season Native Grass Biomass Using Fire, Herbicide, and Nitrogen Applications

Sharon A. Clay, Alexander Smart and David E. Clay

Abstract

The North American Great Plains tallgrass prairie was once a system of native cool and warm season grasses, which have been degraded by non-native invasive plants. Native grass restoration is highly desirable to improve ecosystem functions and productivity. In this two-year study, the impact of fire, herbicide, and nitrogen on productivity and the presence of invasive species [primarily the cool season grass, smooth brome (*Bromus inermis* Leyss.)] and native warm season native grass species [big bluestem (*Andropogon gerardii* Vitman), sideoats and blue grama (*Bouteloua curtipendula* (Michx.) Torr.), and *B. gracilis* (Willd. Ex Kunth) Lag. ex Griffiths] were investigated. Spring fire or a glyphosate application increased warm season grass biomass and decreased cool season grass biomass at peak warm season growth (August) during the treatment year. A second consecutive year of fire or herbicide further increased warm season grass biomass. If left untreated in the second year, cool season grasses tended to increase when sampled in August. Long-term management implementation is needed to suppress the tenacious cool season species and encourage the reestablishment of warm season grass populations.

Keywords: prairie restoration, glyphosate, *Bromus inermis* control, *Andropogon gerardii*

1. Introduction

The tallgrass prairie once covered 170 million acres in North America, but only about 4% remain [1–4]. The tall and mixed grass prairie systems have been in decline since European settlement, with human encroachment, population expansion, and overgrazing with insufficient recovery times contributing to the loss of native prairie acres [3–5]. In addition, the loss of natural disturbances, such as fire and grazing with bison, has decreased native plant presence and diversity in remnant prairie sites [3, 4, 6]. Other threats to this resource include fragmentation, as smaller parcels provide less continuity, and due to more isolated populations, an increase of deleterious genes in a community that reduce fitness [7]. The introduction of non-native species, either purposefully introduced or encroachment from neighboring areas, often dominate the ecosystem [8]. In South Dakota, non-native species are estimated to be present on 82% of its 9.7 million ha of rangeland, and account for at least 25% of the relative canopy cover on 22% of these areas.

Kentucky bluegrass (*Poa pratensis* L.) and smooth brome are typically cited as the most invasive cool season species in the Northern Great Plains, accounting for over 10% of all plant cover and 62% of exotic species cover [9, 10]. Infestations progress from scattered patches to over 90% of a pasture in 30 yrs. [11]. These C_3 species suppress native cool and warm season grass production and can create negative feedback cycles that increasingly perpetuate favorable microclimates for the invasive species, making restoration to natives increasingly difficult [12, 13]. Cool season grasses become dominant by breaking dormancy very early in the season and outcompeting later emerging plants for nutrients, water, and light [13, 14], and later, forming a thick thatch layer that does not readily decompose [12, 13, 15], which slows soil warming, shades the soil surface, and delays the growth of warm season plants [16, 17].

Management practices are needed that reduce invasive species competitiveness and enhance growth and productivity of native species [11, 18]. Spring burns remove thatch layers, allowing up to 40% more sunlight to reach the soil surface [17], which increases soil temperatures, provides more direct light to small plants that could be stunted by shading, and removes or sets back growth of cool-season species, thus reducing competition from larger more vigorous plants. Hulbert [19] reported that thatch removal and exposing the soil surface to light resulted in an increase of warm season species vegetative and reproductive productivity regardless of how the thatch was removed (e.g. clipped or burned). Native warm season grass biomass had greater response to burning (98% biomass increase) than to soil warming (8% biomass increase) [16]. Therefore, prescribed burns often are used to influence prairie species composition.

Timing of prescribed burns can be crucial in influencing species outcome. For example, long-term studies (>54 yrs) of annual Flint Hills (Kansas) burns report that late spring burns (May 1) increased later season biomass of warm-season grasses including big bluestem and Indiangrass (*Sorghastrum nutans* L. Nash), whereas sedges and perennial forb biomass decreased, as the burn time coincided with their emergence [20, 21]. However, early spring (March 20) and winter (December 1) burns favored annual and perennial forbs, and other grass species such as Scribner's panicum [*Dichanthelium oligosanthes* (J.A. Schulte) Gould var. *scribnerianum* (Nash) Gould], and little bluestem (*Schizachyrium scoparium* Michx.) [21]. In a Minnesota study, little bluestem did not change in biomass based on different fire frequency return times (1, 2, or 4 yr intervals) over 27 yr, although Kentucky bluegrass decreased with increasing fire frequency [22]. In South Dakota, Kentucky bluegrass biomass was reduced with either mid-May and mid-June burns, however, big bluestem biomass only increased after the early burn [23]. The dominant species of the starting community (e.g. smooth brome, Kentucky bluegrass, or warm-season grasses) also influences community composition response to different burn timings [24]. Results from these diverse studies indicate that individual plant and community responses to burn timing could assist or hamper land managers in their goal to manipulate species composition.

A second method of influencing grassland composition is adding or withholding nutrients, with or without prescribed burns [24–36]. Burning changes nutrient cycling [26, 27]. Supplemental N application rates and timings can influence species composition and forage production in range and pastures. For example, ammonium nitrate (NH_4NO_3) applied at 10 or 20 g/m^2 was used alone or combined with annual burns [28]. When fire was not used and N was added, forbs increased and exceeded grass production, but had little impact on total overall biomass. In treatments that were burned and received N, C_4 grasses dominated and total biomass increased by 68% [28]. However, the N rate needs to be closely monitored, as anthropogenic N deposition, principally through rainfall and attributed to industrialization, urbanization, and increased fertilizer use [29–31], has doubled the input of available N on the Earth's surface in the recent past [31]. Adding too much N, as little as 2.5 [30] to 10 [32] kg N ha^{-1} $yr.^{-1}$, can reduce native plant species numbers [32] especially those adapted to low soil N [33]. High soil N reservoirs present, especially early in the growing season, may lead to greater non-native plant invasion [34], as these plants tend to acquire and use N more efficiently, thus increasing their biomass and density [35] with a decline of native species growth. Once non-natives establish, a cascade of events may occur that render reversal to the original native community difficult, if not impossible. First, competition for water and other resources further reduces native species density, biomass, and species richness of a site [35]. This is followed by reduction in native species growth, which depletes the native seedbank and, finally, subtle changes of soil biological and chemical properties may occur that hinder repopulation by native plants [36].

Herbicides have also been used, alone and in combination with burns and N application, to control smooth brome and Kentucky bluegrass and manipulate native vegetation restoration [37–41]. Herbicide type, application timing, and rate have been examined in several studies. For example, Bahm et al. [38] used herbicides (imazapic, imazapyr, sulfosulfuron) alone or in combination with spring and fall applications to target smooth brome and Kentucky bluegrass. Herbicide treatments were compared to a fall burn or non-treated control. They reported that burn alone did not reduce either grass species, but spring or fall applications of most herbicides reduced smooth brome from 64% cover to 10% cover after 3 years of treatment. Although Kentucky bluegrass was more recalcitrant than smooth brome in several studies [38, 39], native grass cover or species richness increased in treatments. Others have reported greater control of smooth brome with atrazine applied in spring [40] although native grasses were less injured with a fall application of glyphosate [41].

Prescribed burns with modest applications of fertilizer have been shown to sustain and increase native grass species biomass and forage production, improve wildlife habitat, and decrease the need for weed management in more southern, western, and northern U.S. regions [8, 20–34, 40–41]. However, fewer studies have been performed in the Northern Great Plains [38, 39] where soils, plant composition, rainfall, and temperatures widely differ. The objective of this study was to examine the influence of combinations of burn, herbicide application (fall or spring), and N application (spring, summer, or fall) on non-native cool season grass competition and biomass, and native warm season grass biomass.

2. Using fire, herbicide, and nitrogen to manipulate cool and warm season grasses

2.1 Methods and materials

2.1.1 Site description

Two sites were used in this experiment, one in far-eastern South Dakota at Volga (N 44°23′1.53“, W 96°57’29.39”) and a more western site (100 km west and 50 km south of Volga) at Artesian (N 44°5.80′, W 97°54.56′). Both sites have hot summers with temperatures exceeding 32°C, periodic summer droughts, and cold (air temperatures as low as −40°C) snowy winters.

The Artesian site, located in Sanborn County was used for cut hay and pasture. The soil type was predominantly Houdek-Dudley complex (fine-loamy, mixed Typic Argiustolls and fine, mixed Typic Natrustolls) with a 0–9 percent slope [42] and is glacial till loamy claypan with thin uplands and wet meadows. The plots were in upland positions. The dominant species were Kentucky bluegrass and smooth brome, with big bluestem, sideoats grama, and little bluestem present in plot area.

The Volga site, in Brookings County, was located on a heavily cool-season grass infested remnant prairie surrounded by pasturelands and was rotationally grazed. The soil mapping unit was a Buse-Poinsett complex (fine-loamy, mixed Typic Calciudolls and fine-silty, Calcic Hapludolls) [42] that has an ecological description as a thin loamy soil. Plots were located at the summit and shoulder-slope positions in an area with a 2–10 percent slope, although shallow marshes and wet meadows were also present at the site. Dominant grass species prior to treatment included smooth brome and Kentucky bluegrass. Big bluestem and sideoats grama were present, but not abundant, in the pretreatment survey of vegetative composition.

2.1.2 Experimental design and treatments

The experimental design was a randomized split-block (12 by 6 m) split-plot with four replications. The main effect treatments were fire, N, and herbicide application with each main plot split into four subplots (6 by 3 m). Plot areas were selected and established in fall 2009 (Location 1) and fall 2010 (Location 2). Location 1 plots were treated in the fall of 2009 with herbicide or N in the appropriate plots, and then in spring 2010 with the remaining treatments (designated as YR1). Location 1 plots at the Artesian site were subdivided in fall 2010 with half the plot 'recovering' (RECOVERY) from the 2010 treatment (i.e. no further treatment), and the other half treated (in fall 2010, when appropriate, and spring 2011) for a second year with the same treatment (YR2). Location 2 plots were a repeat in time of the YR1 treatments (Volga and Artesian).

2.1.3 Fire treatments

The two main fire treatments were 1) not burned or 2) burn in the spring. The fire main plots were then divided into four subplots with treatments of N: (1) no N applied; or N applied at 25 kg N ha^{-1} as NH_4NO_3 (2) in October prior to burn; (3) in April prior to burn, to stimulate cool season grass growth to maximize injury, or (4) in June after the burn to stimulate warm season species growth.

The spring burn was conducted between April 21 to May 9 (2010 and 2011) and depended on vegetative development, dryness, and wind speed for safety. The fires were started with a drip-torch and back burned or using a weed burner. Water was applied around the plot edges, and a fire crew was present on plot perimeters with spray equipment to contain the fire within a plot area. Thatch depths at Volga averaged 22 cm (±6), and 15 cm (± 1) in 2010 and 2011, respectively, just prior to the burn. At Artesian, thatch depths were 4 cm (±2) (2010 YR1), 13 cm (±3) (2011 YR2), and 15 cm (±3) (2011 YR1). Fire temperatures [monitored by placing four slides painted with Tempilaq thermopaints (LA-CO Industries, Elk Grove, IL, USA) at the soil surface per plot] ranged from 79 to 343°C (average = 219°C) at Volga and from 79 to 325°C (average = 198°C) at Artesian.

2.1.4 Nitrogen treatments (no fire/no herbicide)

The nitrogen block was split into four treatments. The treatments were no N, 25 kg N ha^{-1} as NH_4NO_3 applied in April or June, and a double treatment, 25 kg N ha^{-1} applied first in October and then the following April.

2.1.5 Herbicide treatments

The herbicide treatment was glyphosate (applied with ammonium sulfate and nonionic surfactant). The main herbicide treatment plot was divided into subplots based on times of herbicide and N application (base rate of 25 kg N ha^{-1}). The treatments were herbicide October/no N; herbicide May/no N; April N followed by (F/B) herbicide May; herbicide May F/B June N. The October glyphosate rate was 1.5 kg ai ha^{-1} applied after warm season grass senescence but prior to killing frost. The April/May rate was 0.38 kg ai ha^{-1} applied after cool season grass emergence, but prior to emergence of warm season grasses.

2.1.6 Data collection

At both sites, baseline vegetative biomass was collected during the peak of warm-season growth (late August/early September) in 2009 prior to treatment application. Vegetation samples were cut at soil surface in three 0.25 m^2 quadrats per main plot. Samples were dried at 32°C until constant weight and separated into functional groups of native grasses, non-native species (mostly grasses), and forbs, and weighed.

After spring treatments, visual cover assessments of warm season grass, cool season grass, and forb were evaluated twice during 2010 and 2011: at the peak of cool season grass growth (mid/late June) and at the peak of warm season grass growth (late August/early September). The percentages of bare-ground and litter were estimated. At the Artesian site, all plots [YR1, Location 1 and 2; YR2 and Recovery (Location 1)] were sampled with vegetation cut from a 0.25 m^2 quadrat after both cool and warm season grass peak growth assessments. At the Volga site, species cover at peak cool season grass development was visually assessed. Vegetative sampling occurred in the YR1 Location 1 and 2 plots at peak warm season growth in August of each year. These vegetation samples were dried, separated into functional groups as described above, and weighed.

2.1.7 Statistical analysis

PROC MIXED and PROC GLM [43] were used to analyze the data by site, location, and sampling date to calculate the difference of least mean square and determine differences among treatments. To determine if treatments over time accelerated warm season grass growth and decreased cool season non-native grass growth, repeated measure ANOVA was used in SAS with a PROC MIXED statement. This analysis helped determine variation among samples, variation among sample timing, and residual variation [44]. The repeated measures design uses the control from each sampled subject and has been called within-subjects ANOVA or randomized-blocks ANOVA [45]. The null hypothesis for this study was H_0: $\mu_1 = \mu_2 = \ldots = \mu_i$, i.e. treatments would not influence estimated cover or vegetative biomass compared with the untreated control.

2.2 Results

2.2.1 Climate

2.2.1.1 Artesian

In 2009, prior to the October N treatment, the average temperatures for the growing season were about 2°C below the 30 yr (1971–2001) average, whereas precipitation ranged from 4 to 8 mm above the 30 yr average for June, July, and August [46]. Growing degree days (base 10°C) from 15 April to 15 September were 5% above the 30 yr normal of 1240 GDD in 2010 and 5% below normal for 2011. Precipitation from January through August totaled 483, 940, and 454 mm for 2009, 2010, and 2011, respectively, compared with the 30 yr average of 428 mm.

2.2.1.2 Volga

Temperatures and rainfall for 2009 were near the 30 yr average of 1154 GDD and 439 mm, respectively [46]. The 2010 and 2011 seasons were warmer and wetter than the 30 yr average. GDD for 15 April to 15 September for 2010 and 2011 were about 10% above the 30 yr average each year. January through August precipitation for 2010 and 2011 was much greater than the 30 yr average and totaled 720 (+64%) and 568 (+30%) mm, respectively.

2.2.2 Treatment response - artesian

2.2.2.1 Baseline sampling

In August 2009 during visual assessment at warm season peak growth, warm and cool season grass canopy covers were similar, each occupying about 50% of the canopy. Biomasses of these functional groups were similar and averaged 145 g m^{-2} (± 82) for warm season grasses and 179 g m^{-2} (± 88) for cool season grasses. In April 2010 before spring treatments, baseline visual cover assessments were similar among blocks, with warm season and cool season grass covers estimated at 30 and 65%, respectively. However, warm and cool season grass biomasses were similar and averaged 123 (± 73) and 96 g m^{-2} (± 62), respectively.

2.2.2.2 One year of treatment

Due to differences in precipitation between 2010 and 2011 (i.e. 2010 had 150% more rainfall than 2011), Location 1 and 2 data were analyzed by year. The cool and warm season grass biomass in late June 2010 (peak cool season biomass) averaged 181 (± 74) and 41 (± 9) g m^{-2} in control plots (**Figure 1A**). Fire alone and April N FB fire reduced cool season biomass to about 30 (±6) g m^{-2}, whereas cool season biomass averaged 107 (±12) g m^{-2} in the October N FB fire and fire FB mid-June N treatments (**Figure 1A**). A visual example of the plots pre- and post-fire is provided in **Figure 2**. Glyphosate applied in May alone or FB June N reduced cool season biomass. Warm season biomass was greater than the control in all glyphosate treatments, expect when FB June N application. Nitrogen treatments applied in April or June N averaged 388 (±25) g m^{-2} of cool season grass biomass, twice as much compared with control plots.

At the peak of warm season grass (August 2010) (**Figure 1B**), warm season and cool season grass biomass in control plots averaged 173 and 149 g m^{-2}, respectively. Fire alone or in combination with any N treatment increased warm season grass biomass by at least 120% and decreased cool season biomass by about 50%. The April N FB glyphosate had the greatest warm season biomass of any treatment and averaged over 450 g m^{-2}. Other herbicide treatments had warm and cool season grass biomass that was similar to the control. The June N treatment increased both warm and cool season biomass and April N had similar warm season biomass to the control but also increased cool season grass biomass.

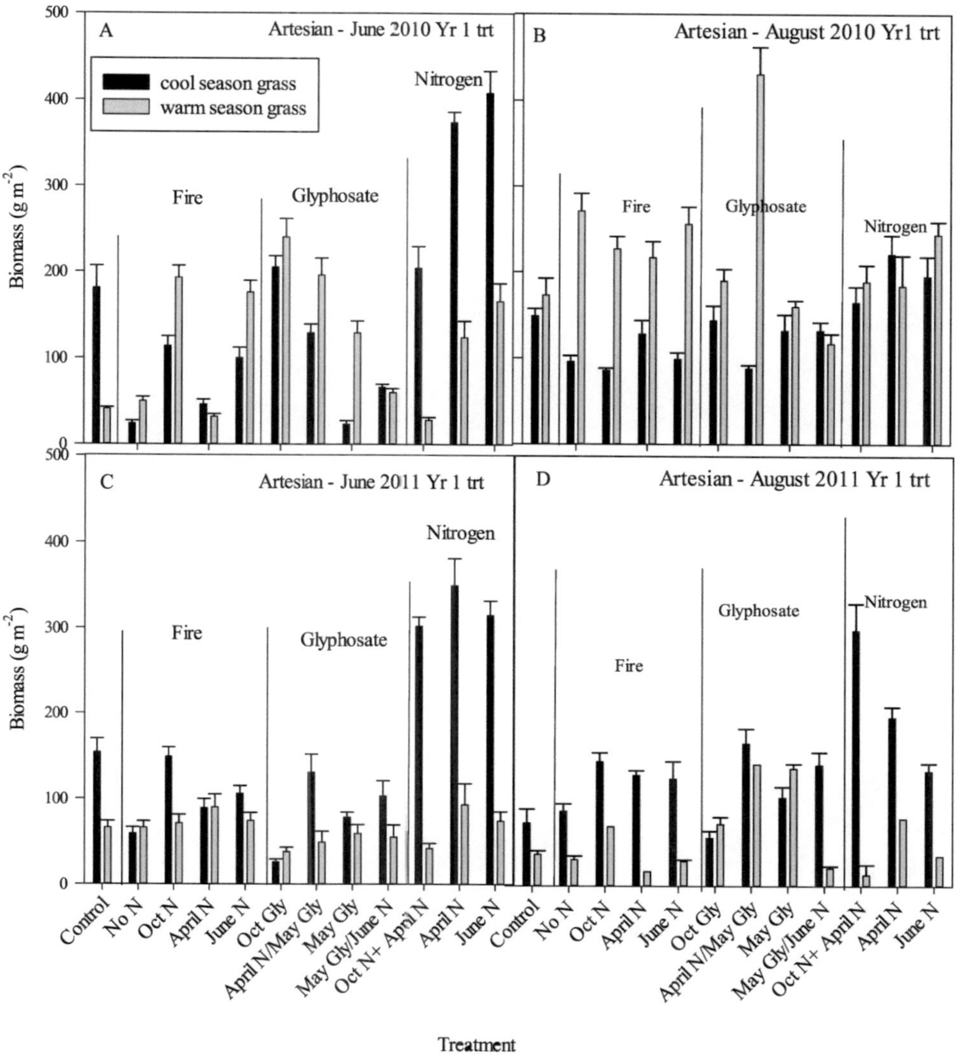

Figure 1.
Cool and warm season grass biomass by treatment and year in June [(A) 2010; (C) 2011] and August [(B) 2010; (D) 2011] samplings after a single year of treatment at Artesian, SD.

In June 2011, the fire treatments reduced cool season grass biomass, except when N was applied the previous October (**Figure 1C**). Glyphosate treatments, except for April N FB glyphosate, also had a cool season biomass reduction. The N treatments alone, however, stimulated cool season grass growth and did not stimulate growth of warm season species. Unlike 2010 when warm season growth showed an increase at the August sampling with many treatments, most treatments had greater cool season biomass with all N treatments having the greatest amounts (**Figure 1D**). The yearly differences between the treatments were partly due to the lower rainfall in 2011 and a heavy infestation of the biennial, sweet clover [*Melilotus officinalis* (L.) Lam.], which bolted and shaded out both native and non-native grasses throughout the entire location.

2.2.2.3 Two years of treatment

After two consecutive years of fire treatment at location 1, when sampled in August (peak of warm season species growth) (**Figure 3A**), October N F/B spring

Figure 2.
Example of a fire plot at Artesian, SD, (A) Prefire in April, with cool season invasive grasses present (B) during fire treatment, (C) about two weeks post-fire, (D) native warm season grass growth, August sampling, 2010.

fire had the least cool season grass biomass, (reduced by 37%) and the greatest warm season biomass. Glyphosate applied in May had the greatest warm season grass biomass of all the treatments. Applying N in April FB May glyphosate resulted in less warm season grass biomass. When N followed glyphosate application, cool season grasses, rather than warm season, appeared to be stimulated. Nitrogen alone resulted in greater cool season grass biomass even in August.

2.2.2.4 Recovery after one year of treatment

In the June sampling, most plots had high amounts of cool season grass and low warm season grass presence (**Figure 3B**). In the August sampling, warm season grasses in all fire plots, except fire FB June N, had greater warm season and less cool season grasses (**Figure 3C**). The glyphosate and nitrogen treatments had

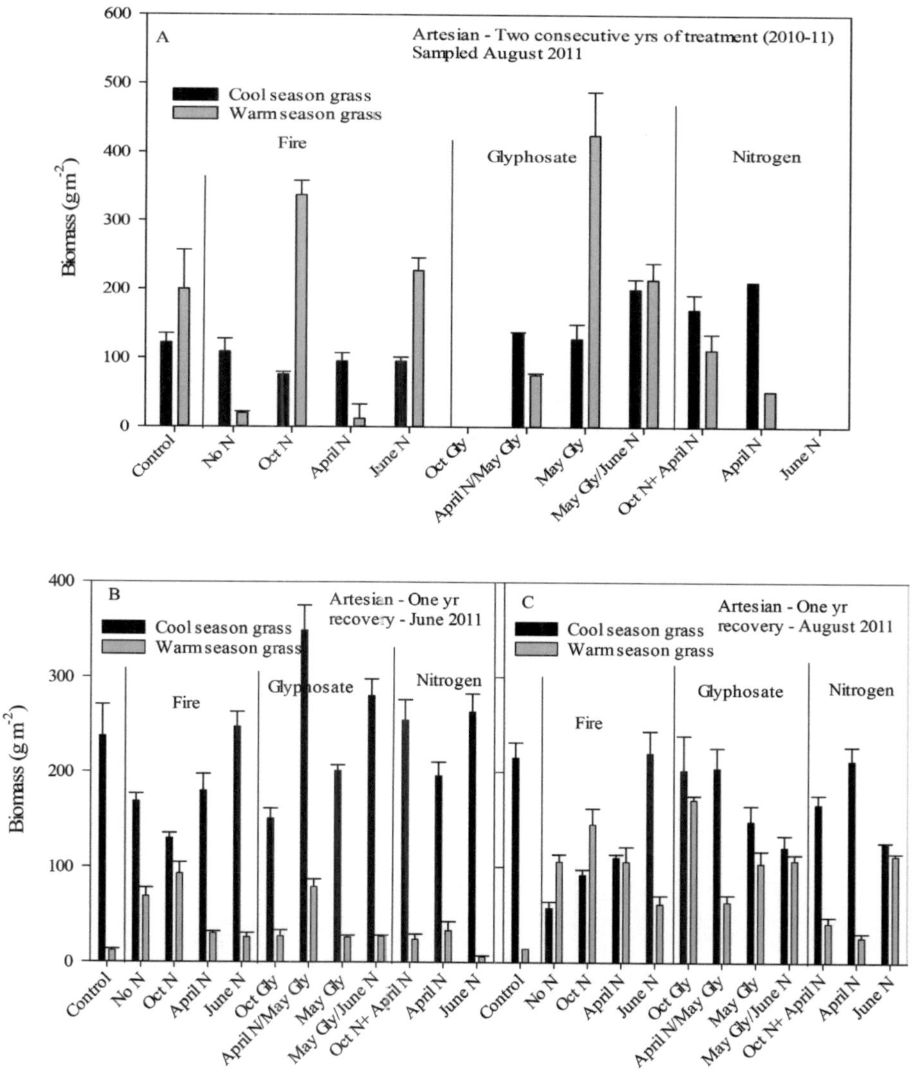

Figure 3.
Cool and warm season grass biomass in August sampling after two consecutive years of treatment (A), and in recovery plots (treated in 2010) for June (B) and August (C) of 2011.

greater warm season grass biomass than the control, but little reduction in cool season grass biomass.

2.2.3 Treatment response - Volga

2.2.3.1 Baseline data

Baseline data collected in the fall of 2009 (for Location 1 plots) and 2010 (for Location 2 plots) had visual cover estimations for warm season grass species at 35%, cool season grass species 65%, and 5% forbs.

2.2.3.2 One year of treatment

At location 1 (2010 treatments), visual assessments of canopy cover were similar among the control and all treatments at the June sampling date (data not shown).

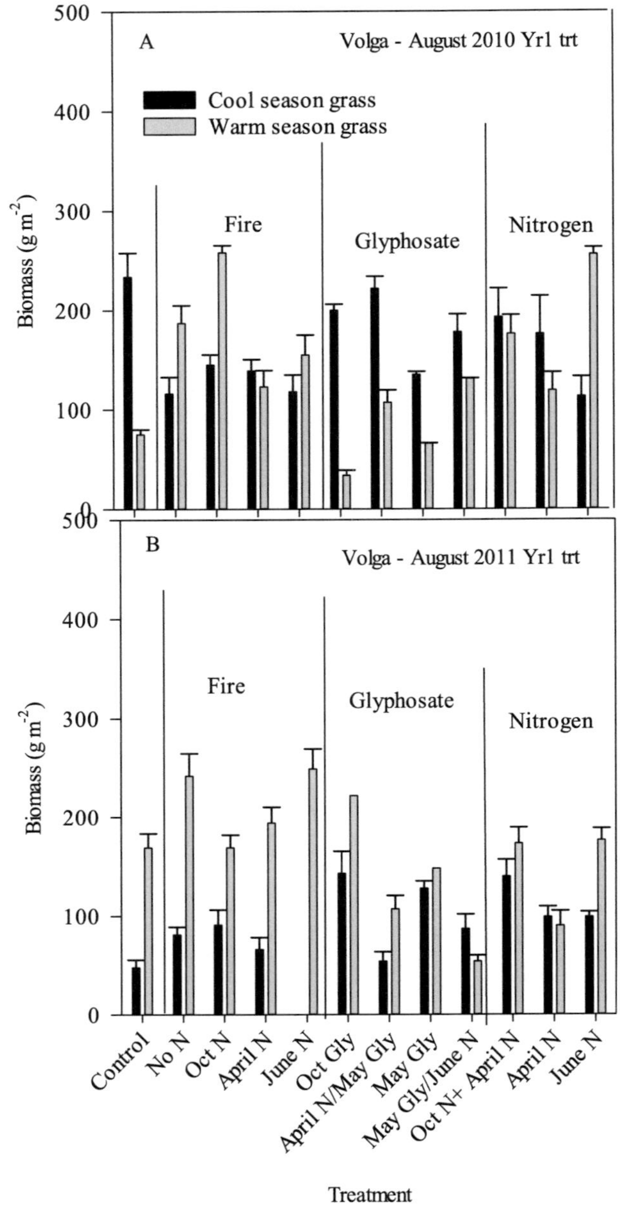

Figure 4.
Cool and warm season grass biomass by treatment sampled in August of 2010 and 2011 after the first year treatment at Volga, SD site.

Warm season grass biomass was 50 to 150% greater in fire treatments and had better cool season grass control in 2011 than 2010 (**Figure 4A** and **B**). Warm season grass biomass in the glyphosate and N treatments were similar to the control. Cool season grass biomass at the August 2010 sampling was reduced by all fire treatments, but all other treatments in 2010 and all treatments in 2011 had cool season grass biomass that was almost equal to the control.

2.2.3.3 Two years of treatment and recovery

After two years of treatment, visual cover assessments in all treatments were similar to the control. Biomass differences from the control were seen as an increase in warm-season grass species when October N was followed by the spring fire treatment (+1613%, $P < 0.05$), although cool season grass biomass did not differ from the control. Recovery plots at Volga did not show any differences from the control for either estimated cover or biomass.

3. Discussion

The introduction of cool season non-native aggressive grasses to the Northern Great Plains has reduced the warm season native grass component of the tallgrass prairie. In this study, fire, herbicide, and nitrogen treatments, alone and in combination with different application timings were used to examine the response of both the cool season invasive species and the warm season desired species with one year of treatment, two consecutive years of treatment, and the recovery response of the grasses if only treated one year. At both Artesian and Volga, the dominant native species was big bluestem, with lesser amounts of sideoats grama, and blue grama. These species have been reported to benefit from late spring burns [21] and our study agrees with those findings.

Prescribed burns removed the litter layer, allowed soil warming, and increased light reaching the soil surface, with concomitant increases in plant growth, all reported to increase warm season species growth [16]. In three out of the four first year plots, spring burns with or without N increased warm season species biomass as observed in August of the treatment year. The exception was Artesian 2011 first year plots, when sweet clover dominated all plots at both the June and August samplings. April N FB glyphosate increased warm season grass biomass in August 2010 but this response was not as pronounced in 2011. Warm season grasses at the both sites benefited from the two consecutive years of fire, but in Artesian, October N FB fire or fire FB June N was best, whereas at Volga, fire with no N was the best treatment. However, because litter depth is typically less after the first year of fire, lethal heat (>60°C) [47] at the soil surface tends to be longer if fire return times are biennial rather than annual [48]. Glyphosate applied for two consecutive years also aided in warm season grass growth. However, N alone would not be recommended, as cool season grasses dominated in all these plot areas. Recovery plots treated only with fire had increased warm season species biomass compared with control areas at both sites. Nitrogen application either before or after the fire treatment did not benefit warm season grass growth.

Vinton and Goergen [49] reported that smooth brome has a competitive advantage over native tallgrass species due to increased efficiency to cycle N. This characteristic may increase the persistence of smooth brome due to N deposition due to anthropogenic factors. Throughout this experiment, the application of N alone at any timing increased cool season grass biomass.

We were hopeful that cool season species would be more prone to fire or herbicide injury following either October or April N application. Increased injury to the cool season species (as observed by decreases in cool season biomass) only was observed in a few select treatments. In addition, the N added after fire or herbicide was, in theory, supposed to invigorate the warm season grass growth, as cool season grass growth should have been slowed, leaving the N for the late emerging species. However, adding N in June had limited success in increasing warm season grass growth, as observed at the Volga site in both years.

4. Implications and conclusions

Warm season grass growth had the most consistent positive response to fire, which also helped reduce the cool season non-native grass species. The application of fire, however, is not without risk, needs to be carefully utilized, and only applied when weather conditions are optimal and with trained personnel. There may be other considerations when using prescribed burns [50] and local fire officials, neighbors, and state agencies may need to be consulted and notified the day of the burn [51]. Nevertheless, prescribed burns may provide land managers and producers a simple and inexpensive way to reinvigorate the tallgrass prairie in eastern South Dakota. Location, soil type, species present, land use history, timing, temperature, and duration of fire [21–23, 48, 52], and the presence and number of active buds on the perennial grass tillers [24, 53, 54] are components that influence the restoration of areas to warm season native grasses, or maintain the continued presence of the cool season invasive species. Two years of consecutive treatment was more effective for warm season grass restoration than a single year, as seen by the rapid return of cool season grasses in the recovery plots. In addition, at times, a setup fall treatment of nitrogen or non-selective herbicide applied prior to the spring treatment may benefit warm season grass response, although these management options need further study.

Acknowledgements

Partial funding provided by SDSU Ag Experiment Station, NCR-SARE – 2010 Graduate Student Award, and USDA/CSREES Grassland grant #2008-38415-19596. Project field work was conducted by former master degree graduate student Ms. Shauna Waughtel-Johnson. Thanks to Dr. Michelle Ohrtman for help with data analysis.

Author details

Sharon A. Clay[1]*, Alexander Smart[2] and David E. Clay[1]

1 Department of Agronomy, Horticulture, and Plant Science, South Dakota State University, Brookings, SD, USA

2 Department of Natural Resource Management, South Dakota State University, Brookings, SD, USA

*Address all correspondence to: sharon.clay@sdstate.edu

References

[1] US Park Service [Internet]. 2019. Available from: https://www.nps.gov/tapr/index.htm [Accessed: 01 March 2019]

[2] Blankenship B, Bradshaw M, Hinkle M, Schaub R [Internet]. 2019. What Are Some of the Thinkgs That Threaten Prairies and Contribute to Their Loss or Destruction? Available from: https://www.asc.ohio-state.edu/accad/womenandtech/2004/research%20pages/Loss/loss.html [Accessed: 15 February 2019]

[3] Robertson KR, Anderson RC, Schwartz MW. The tallgrass prairie mosaic. In: Schwartz MW, editor. Conservation in Highly Fragmented Landscapes. Boston, MA: Springer; 1997

[4] Risser PG, Bimey EC, Blocker HD, May SW, Parton WJ, Wiens JA. The True Prairie Ecosystem. Stroudsburg, Pennsylvania: Hutchinson Ross; 1981

[5] Albertson FW. Man's disorder of nature's design in the great plains. Transactions of the Kansas Academy of Science. 1949;**52**:117-131

[6] Howe HF. Managing species-diversity in tallgrass prairie-assumptions and implications. Conservation Biology. 1994;**8**:691-704

[7] Mannouris C, Byers DL. The impact of habitat fragmentation on fitness-related traits in a native prairie plant, *Chamaecrista fasciculata* (Fabaceae). Biological Journal of the Linnean Society. 2013;**108**:55-67

[8] Bock JH, Bock CE. Tallgrass prairie: Remnants and relicts. Great Plains Research. 1998;**8**:213-230

[9] [USDA] U.S. Department of Agriculture 2018 National Resources Inventory Rangeland Resource Assessment. Natural Resources Conservation Service. Washington, DC. Available from: https://www.nrcs.usda.gov/wps/portal/nrcs/detail/national/technical/nra/nri/results/?cid=nrcseprd1343028 [Accessed: 15 March 2019]

[10] Cully AC, Cully JF Jr, Hiebert RD. Invasion of exotic plant species in tallgrass prairie fragments. Conservation Biology. 2003;**17**:990-998

[11] Grant TA, Flanders-Wanner B, Shaffer TL, Murphy RK, Knutsen GA. An emerging crisis across northern prairie refuges: Prevalence of invasive plants and a plan for adaptive management. Ecological Restoration. 2009;**27**:58-65

[12] Sather N. Element Stewardship Abstract for *Bromus inermis*. Arlington, VA: The Nature Conservancy; 1987. Available from: http://www.invasive.org/weedcd/pdfs/tncweeds/bromine.pdf [Accessed: 05 March 2019]

[13] Toledo D, Sanderson M, Spaeth K, Herndrickson J, Printz J. Extent of Kentucky bluegrass and its effect on native plant species diversity and ecosystem services in the northern great plains of the United States. Invasive Plant Science and Management. 2014;**7**:543-552

[14] Benson EJ, Hartnett DC. The role of seed and vegetative reproduction in plant recruitment and demography in tallgrass prairie. Plant Ecology. 2006;**187**:163-177

[15] Hendrickson JR, Lund C. Plant community and target species affect responses to restoration strategies. Rangeland Ecology & Management. 2010;**63**:435-442

[16] Hulbert LC. Causes of fire effect in tallgrass prairie. Ecology. 1988;**69**:46-58

[17] Knapp AK. Post-burn differences in solar radiation, leaf temperature and water stress influencing production in a lowland tallgrass prairie. American Journal of Botany. 1984;**71**:220-227

[18] Taylor RV, Pokorny ML, Mangold J, Rudd N. Can a combination of grazing, herbicides, and seeding facilitate succession in old fields? Ecological Restoration. 2013;**31**:141-143

[19] Hulbert LC. Fire and litter effects in undisturbed bluestem prairie in Kansas. Ecology. 1969;**50**:874-877

[20] McMurphy WE, Anderson KL. Burning Flint Hills Range. Journal of Range Management. 1965;**18**:265-269

[21] Towne G, Owensby C. Long-term effects of annual burning at different dates in ungrazed Kansas tallgrass prairie. Journal of Range Management. 1984;**37**:392-397

[22] Li W, Zuo X, Knops JMH. Different fire frequency impacts over 27 years on vegetation succession in an infertile old-field grassland. Rangeland Ecology & Management. 2013;**66**:267-273

[23] Engle DM, Bultsma PM. Burning of northern mixed prairie during drought. Journal of Range Management. 1984;**34**:398-401

[24] Hendrickson JR, Lund C. Plant communities and target species affect responses to restoration strategies. Rangeland Ecology & Management. 2010;**63**:435-442

[25] Boughton EG, Bohlen PJ, Maki JH. Effects of experimental season of prescribed fire and nutrient addition on structure and function of previously grazed grassland. Journal of Plant Ecology. 2017;**11**:576-584

[26] Turner CL, Blair JM, Schartz RJ, Neel JC. Soil N and plant response to fire, topography, and supplemental N in tallgrass prairie. Ecology. 1997;**78**:1832-1843

[27] Risser PG, Parton WJ. Ecosystems of the tallgrass prairie: Nitrogen cycle. Ecology. 1982;**65**:1342-1351

[28] Seastedt TR, Briggs JM, Gibson DJ. Controls of nitrogen limitation in tallgrass prairie. Oecologia. 1991;**87**:72-79

[29] Wedin D, Tilman D. Competition among grasses along a nitrogen gradient: Initial conditions and mechanism of competition. Ecological Monographs. 1993;**63**:199-229

[30] Stevens CJ, Dise NB, Mountford JO, Gowing DJ. Impact of nitrogen deposition on the species richness of grasslands. Science. 2004;**303**:1876-1879

[31] Vitousek PM, Aber JD, Howarth RW, Likens GE, Matson PA, Schindler DW, et al. Human alteration of the global nitrogen cycle: Sources and consequences. Ecological Applications. 1997;**7**:737-750

[32] Clark CM, Tilman D. Loss of plant species after chronic low-level nitrogen deposition to prairie grasslands. Nature. 2008;**451**:712-715

[33] Gotelli NJ, Ellison AM. Nitrogen deposition and extinction risk in the northern pitch plant, *Sarracenia purpurea*. Ecology. 2002;**83**:2758-2765

[34] Vasquez E, Sheley R, Svejcar T. Creating invasion resistant soils via nitrogen management. Invasive Plant Science and Management. 2008;**1**:304-314

[35] Brooks ML. Effects of increased soil nitrogen on the dominance of alien annual plants in the Mojave desert. Journal of Applied Ecology. 2003;**40**:344-353

[36] Estrella S, Kneitel JM. Invasion age and invader removal alter species cover and composition at the Suisun tidal marsh, California, USA. Diversity. 2011;**3**:235-251

[37] Gillen RL, Rollins D, Stritzke JF. Atrazine, spring burning, and nitrogen for improvement of tallgrass prairie. Journal of Range Management. 1987;**40**:444-447

[38] Bahm MA, Barnes TG, Jensen KC. Herbicide and fire effects on smooth brome (*Bromus inermis*) and Kentucky bluegrass (*Poa pratensis*) in invaded prairie remnants. Invasive Plant Science and Management. 2011;**4**:189-197

[39] Link A, Kobiela B, DeKeyser S, Huffington M. Effectiveness of burning herbicide, and seeding toward restoring rangelands in southeastern North Dakota. Rangeland Ecology & Management. 2017;**70**:599-603. DOI: 10.1016/j.rama.2017.03.001

[40] Willson GD, Stubbendieck J. Suppression of smooth brome by atrazine, mowing, and fire. The Prairie Naturalist. 1996;**28**:13-20

[41] Anderson B. Converting smooth brome pasture to warm-season grasses. In: Proceedings of the Thirteenth North American Prairie Conference, Windsor, Ontario. 1994. pp. 157-160

[42] NRCS, USDA. 2019. Web Soil Survey. Soil Series Classification: View by Name. Available from: http://websoilsurvey.nrcs.usda.gov/app/WebSoilSurvey.aspx [Accessed: 25 January 2019]

[43] SAS Institute Inc. SAS Software for Windows, v. 9.2. Cary, NC: SAS Institute, Inc.; 2008

[44] Hinkle DE, Wiersma W, Jurs SG. Applied Statistics for the Behavioral Sciences. 5th ed. Boston, MA: Houghton Mifflin Company; 2003

[45] Vogt WP. Dictionary of Statistics and Methodology: A Non-Technical Guide for the Social Sciences. 3rd ed. Thousand Oaks, CA: Sage Publications. 2005. pp. 259-268

[46] High Plains Regional Climate Center [Internet] Brookings (2NE, ID 391076) and Forestburg (3NE ID 393029), South Dakota Climate Summary 2011. Available from: https://hprcc.unl.edu/ [Accessed: 16 September 2012]

[47] Vermeire LT, Rinella MJ. Fire alters emergence of invasive plant species from soil surface deposited seeds. Weed Science. 2009;**57**:304-310

[48] Ohrtman MK, Clay SA, Smart AJ. Surface termperatures and durations associated with spring prescribed fires in eastern South Dakota tallgrass prairies. American Midland Naturalist. 2015;**173**:88-98

[49] Vinton MA, Goergen EM. Plant-soil feedbacks contribute to the persistence of *Bromus inermis* in tallgrass prairie. Ecosystems. 2006;**9**:967-976

[50] Randall J, Harr R. Considerations for Prescribed Burning: Ignition Techniques. Iowa State University Extension. 2012. PMR2088E. Available from: https://www.extension.iastate.edu/forestry/publications/pdf_files/pm2088e.pdf [Accessed: 20 March 2019]

[51] Porter M. How to Conduct a Prescribed Burn. 2000. Available from: https://www.noble.org/news/publications/ag-news-and-views/2000/february/how-to-conduct-a-prescribed-burn/ [Accessed: 15 March 2019]

[52] Vermeire LT, Russell ML. Seasonal timing of fire alters biomass and species composition of northern mixed prairie. Rangeland Ecology & Management. 2018;**71**:714-720

[53] Russell ML, Vermeire LT, Ganguli AC, Hendrickson JR. Fire return interval and season of fire alter bud banks. Rangeland Ecology & Management. 2019;**72**:542-550

[54] Ott JP, Butler JL, Rong Y, Xu L. Greater bud outgrowth of *Bromus inermis* than *Pascopyrum smithii* under multiple environmental conditions. Journal of Plant Ecology. 2017;**10**:518-527

Chapter 5

Effects of Grazing Intensity on Belowground Carbon and Nitrogen Cycling

Guiyao Zhou, Lingyan Zhou and Xuhui Zhou

Abstract

Livestock grazing activities substantially affect grassland ecosystem functions such as carbon (C) and nitrogen (N) cycles. Although numerous individual and synthesized studies had been conducted, how grazing, especially its intensity, affects belowground C and N cycling in grasslands remains poorly understood. In this chapter, our previous published studies were summarized to elucidate the 19 variables associated with belowground C and N cycling in response to livestock grazing across global grasslands. Overall, grazing significantly decreased belowground C and N pools in grassland ecosystems, with the largest decreases observed in microbial biomass C and N (21.62 and 24.40%, respectively). However, the response magnitude and directions of belowground C- and N-related variables largely depend on grazing intensities. Specifically, light grazing promoted soil C and N sequestration, whereas moderate and heavy grazing significantly accelerated C and N losses. This study highlights the importance of grazing intensity for belowground C and N cycling, which urges scientists to incorporate it into regional and global models for predicting human disturbance on global grasslands and assessing the climate-biosphere feedbacks accurately.

Keywords: carbon sequestration, CO_2 emission, heavy grazing, mineralization, soil microbial biomass

1. Introduction

The global grasslands cover 59 million km^2 (nearly 40%) of the terrestrial land [1] and store 10–30% of the global soil organic carbon (SOC, [2]). Currently, the majority of grasslands around the world are suffering from overgrazing [3], which may impose profound effects on ecosystem services and functions by altering the biogeochemical cycle, especially on carbon (C) and nitrogen (N) cycles [4, 5]. The altered C and N cycles may lead to a positive or negative climate-biosphere feedback, which in turn amplify or diminish their net effects on biodiversity and stability of grasslands. Therefore, understanding the C and N cycles in response to grazing is crucial for us to better predict future global C balance and enhance the sustainable management of grasslands [6].

Over past 30 years, numerous studies have been conducted to explore the C and N cycles of aboveground processes in response to grazing in grassland ecosystems, which have substantially improved our understanding of the grazing effects and the potential mechanisms [3, 7, 8]. For example, intermediate grazing may increase more aboveground biomass C than light and heavy grazing because of higher plant diversity [9, 10]. However, due to the spatial heterogeneity and methodological difficulties, the effects of grazing on belowground process remain poorly understood especially at global scale. Although plenty of studies have investigated the effects of grazing on belowground C and N cycles, however, diverse results were reported with increase [11], decrease [12] and no changes [13].

Recent studies have found that the contradictory effects of grazing on belowground C and N cycles may be associated with grazing intensities, climatic conditions, and vegetation types [8, 14]. Compared with other factors, grazing intensity may be the key driver regulating belowground C and N cycles because it significantly alters soil microenvironment, soil nutrients availability, plant community structure, and soil microbial diversity [15, 16]. However, current understanding of the effects of grazing intensity on belowground C and N cycles were also contradictory. For example, Schuman et al. [17] found that neither light nor heavy grazing could significantly change total plant biomass and soil C and N pools. By contrast, response of soil C and N pools would decrease with increased grazing intensity in water-limited grassland [18]. These knowledge gaps may trigger great challenges for us to precisely assess the climate-biosphere feedbacks in the future [14]. Therefore, this chapter mainly focused on the general response patterns of belowground C and N cycles to different grazing intensities and explored its underlying mechanisms at the global scale.

2. Belowground C and N pools and fluxes

Grazing intensity significantly affects the belowground C and N pools and fluxes, because grazing intensity alters plant community structure, soil microenvironment, and soil microbial diversity and activity [15, 19, 20]. A meta-analysis of 115 published studies demonstrated that grazing significantly influenced belowground C and N cycles at the global scale (**Figure 1**). Moreover, grazing intensity usually influenced the response magnitude (even direction) of the majority of the assessed belowground C and N pools and fluxes. For example, light grazing increased soil carbon pool (SCP) and soil nitrogen pool (SNP) by 0.78 and 3.24%, respectively ($P < 0.01$, **Figure 2**). However, moderate and heavy grazing significantly decreased SCP by 3.45 and 9.92%, and SNP by 8.41 and 13.04%, respectively, resulting in a diminishing effects on soil C:N ratio from light to heavy grazing (SCN, **Figure 2**). Light grazing may increase the above and belowground biomass, which could stimulate more photosynthetically fixed C allocated to roots and then leading the increase of root exudates and root biomass [12, 14, 21]. Grazing-induced increase in root exudates may further enhance soil C accumulation as well N inputs into soils [22]. Meanwhile, light grazing can also stimulate soil respiration due to the increased root biomass and soil C accumulation [10, 23, 24]. However, both moderate and heavy grazing could markedly decrease SCP and SNP (**Figures 2** and **3**), which was consistent with some previous studies [25–27]. The decreased SCP and SNP may result from that fact that grazing can decrease litter biomass, root C pool and microbial biomass and then lower C inputs to soils (**Figures 3** and **4** [17, 28, 29]).

The altered C and N pools induced by grazing intensity also caused the difference of belowground C and N fluxes. On average, soil respiration (Rs) increased by 11.53% under light intensity, whereas moderate and heavy intensities decreased it by 12.7 and 32.6%, respectively. The weighted response ratios of soil net N mineralization (SNNM) decreased by 48.87–10.85% from light to heavy grazing intensities. However, light grazing did not affect the response ratios of soil net N nitrification (SNNN), but moderate and heavy grazing intensities significantly increased [RR_{++} (SNNN)] by 13.43 and 103.06%, respectively. The differential responses of belowground fluxes may be caused by the following

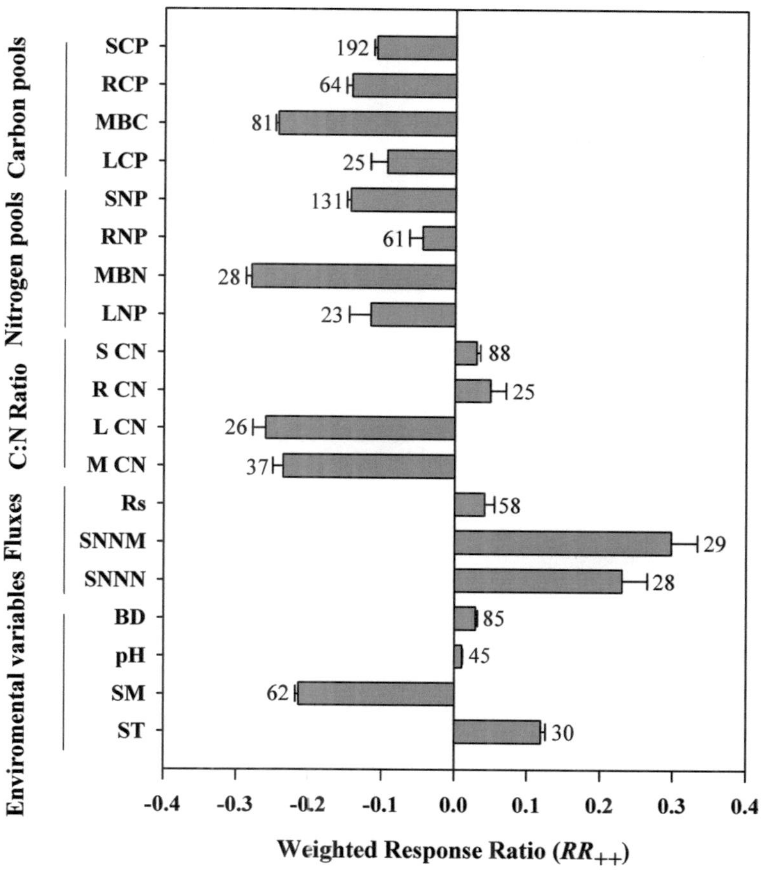

Figure 1.
Weighted response ratio (RR_{++}) of 19 variables related to carbon and nitrogen cycling in response to grazing. Bars represent 95% confidence intervals. The vertical line was drawn at RR_{++} = 0. Numbers for each bar indicate the sample size. SCP, soil carbon pool; RCP, root carbon pool; MBC, microbial biomass carbon; LCP, litter carbon pool; SNP, soil nitrogen pool; RNP, root nitrogen pools; MBN, microbial biomass nitrogen; LNP, litter nitrogen pool; SCN, soil C:N ratio; MCN, microbial biomass C:N ratio; RCN, root C:N ratio; Rs, soil respiration; SNNM, soil net mineralization; SNNN, soil net N nitrification; BD, bulk density; SM, soil moisture; ST, soil temperature.

mechanisms: (1) difference in carbon allocation to roots. The increased C allocation induced by light grazing to root would stimulate the biomass accumulations, which could further increase root activity and C inputs to soil. Moderate and heavy grazing probably also depressed soil infiltrability and nutrient availability, inhibiting plant biomass accumulation and microbial activity [30]; (2) Micro-environment regulations. Light grazing would increase soil moisture because that the enhanced ground covers and decreased soil compaction [21, 31]. Light grazing induced increase in soil temperature and moisture may stimulate plant growth and microbial activities, which would further increase soil respiration [30]. The faster soil evaporation with poor ground cover under moderate and heavy grazing would lower soil moisture, which might also further explain the decreased soil respiration [20, 30, 31].

3. Interaction with biotic and abiotic factors

Grazing effects on belowground carbon and nitrogen cycling were also regulated by biotic (e.g., livestock type) and abiotic factors (e.g., MAP, MAT and soil

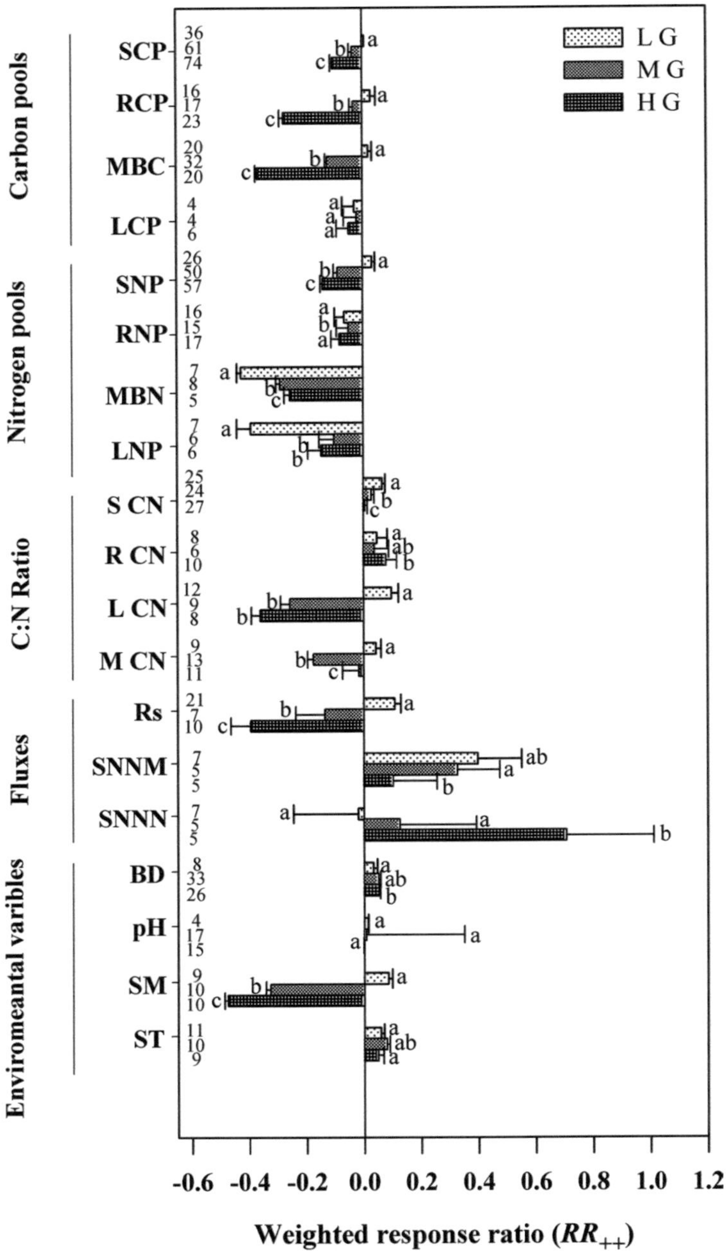

Figure 2.
Weighted response ratio (RR_{++}) of 19 variables related to carbon and nitrogen cycles in response to different grazing intensity. Bars represent RR_{++} ± 95% confidence intervals. The vertical line was drawn at RR_{++} = 0. Numbers for each bar indicate the sample size. Symbols a, b *and* c *represents the significant differences among three grazing intensities for the responses of selected variables to grazing. SCP, soil carbon pools; RCP, root carbon pools; MBC, microbial biomass carbon; LCP, litter carbon pools; SNP, soil nitrogen pools; MBN, microbial biomass nitrogen; LNP, litter nitrogen pools; RNP, root nitrogen pools; SCN, soil C:N ratio; MCN, microbial biomass C:N ratio; RCN, root C:N ratio; Rs, soil respiration; SNNM, soil net mineralization; SNNN, soil net N nitrification; BD, bulk density; SM, soil moisture; ST, soil temperature; LG, light grazing; MG, moderate grazing; HG, heavy grazing.*

depth). It has been showed that root carbon pool (RCP), soil nitrogen pool (SNP) and root nitrogen pool (RNP) decreased more in semi-humid/humid regions (MAP ≥ 400 mm) than in arid/semi-arid regions under grazing (MAP < 400 mm, **Figures 5** and **6**). However, microbial biomass carbon (MBC) and litter carbon pool

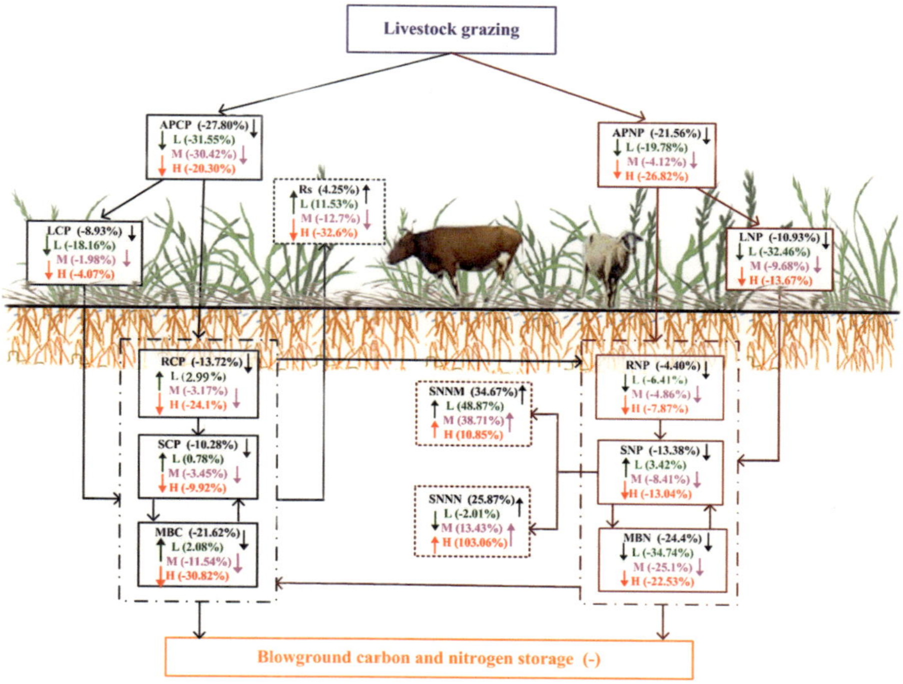

Figure 3.
Potential mechanisms of belowground C and N processes in response to livestock grazing. The numbers refer to percentage change ($e^{RR++} - 1$) × 100% of belowground C and N variables in response to grazing. SCP, soil carbon pools; RCP, root carbon pools; MBC, microbial carbon; LCP, litter carbon pools; SNP, soil nitrogen pools; MBN, microbial nitrogen; LNP, litter nitrogen pools; RNP, root nitrogen pools; APCP, aboveground plant carbon pools; APNP, aboveground plant nitrogen pools. APCP and APNP data was provided in supporting information. L, light grazing intensity presented with green color; M, moderate grazing intensity presented with purplish color; H, heavy grazing intensity presented with red color. ↑, positive response to livestock grazing; ↓, negative response to livestock grazing.

(LCP) exhibited larger negative response to grazing in arid/semi-arid regions than in semi-humid/humid regions. These differences may result from the interactions with precipitation. MAP exhibited a significant positive correlation with the response of SCP ($P < 0.05$), but it was not correlated with response of SNP to grazing (**Figure 7**). Since faster root turnover in wetter regions, grazing lead a larger decrease in RCP in semi-humid/humid than arid/semi-arid climate regions [32]. Due to the close relationship between LCP and MBC, RR(MBC) exhibited a similar response trend with LCP (**Figure 4**, [33]). Grazing significantly decreased MBC and LCP in arid/semi-arid climate, where lower productivity was more responsive to grazing than those in semi-humid/humid conditions. In addition, grazing may substantially reduce MBC in arid/semi-arid climate due to the larger decrease of litter inputs [34, 35]. Furthermore, *Rs* in semi-humid/humid regions increased more than that in arid/semi-arid regions, which might be associated with the existing high net ecosystem productivity [36] and high microbial activity [37] in the wetter regions than those in drier ones. Our study also further found that MAP exhibited a positive correlation with *RR* (*SCP*) (**Figure 7**, **Table 1**), which was consistent with Mcsherry and Ritchie [14] and Hu et al. [38]. Because that plant productivity and microbial activity in wetter areas are usually greater than those in drier regions, the actual responses of SCP to grazing may have been masked, causing weak positive correlation between MAP and SCP [39, 40].

Temperature is another important factor influencing grazing effects on belowground C and N cycles. Our results found that MAT exhibited negative correlations with RR(SCP) and RR(SNP) at global scale (**Figure 7**; **Table 1**). These changes

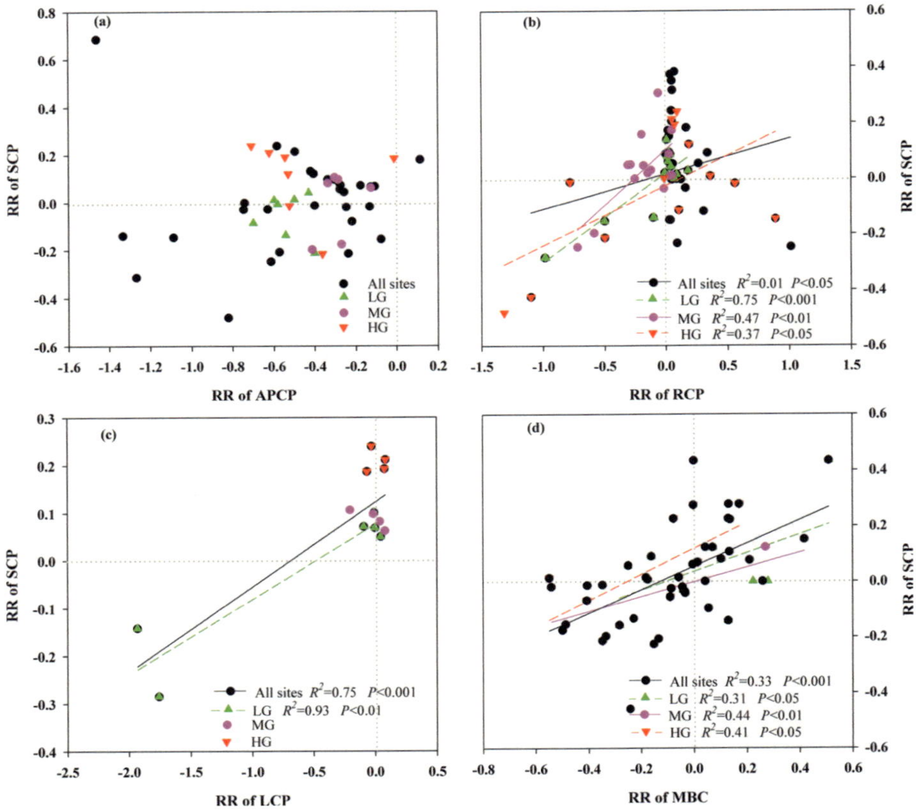

Figure 4.
Relationships of response ratios (RR) of soil carbon pools (SCP) with aboveground carbon pools (APCP, a), root carbon pools (RCP, b), litter carbon pools (LCP, c) and microbial biomass carbon pools (MBC, d). All sites represented the data for all intensities and some with no intensity information—black closed circles; LG, light grazing intensity—green closed triangles; MG, moderate grazing intensity—purple closed circles; HG, heavy grazing intensity—red closed triangles.

may result from the fact that grasslands with higher MAT in tropical and temperate regions usually have greater microbial activity than those in boreal regions with the lower MAT [32]. The higher microbial activity in high-MAT regions can usually accelerate decomposition of soil organic matter and increase turnover rate, and then decrease SCP and SNP more in those grazed ecosystems, resulting in the negative correlation between MAT and *RR* (*SCP*) or *RR* (*SNP*). On the other hand, soil temperature, water content and their interactions fundamentally determine the temporal dynamics of C cycle in grassland ecosystem, especially for soil respiration [41].

Different livestock types and soil depths showed different magnitudes of changes (even direction) for many of the considered variables (**Figure 6**). Using meta-analysis, we found that sheep grazing induced the changes in SCP, SNP, RCP and RNP exhibited a greater decrease than those by cattle. These changes may result from the difference in foraging selectivity by different livestock, causing the variation of plant species composition and community structure, which further induced the difference of C and N inputs/outputs [29]. We also found that the response of MBC at the depth of <15 cm to grazing was positive, while this at depth of >15 cm was negative. Grazing may induced the spatial variations of root distribution and sensitivity to environment within plant–soil system at different depths, which thus causing the different response of belowground C and N cycles to grazing activity [17, 42, 43].

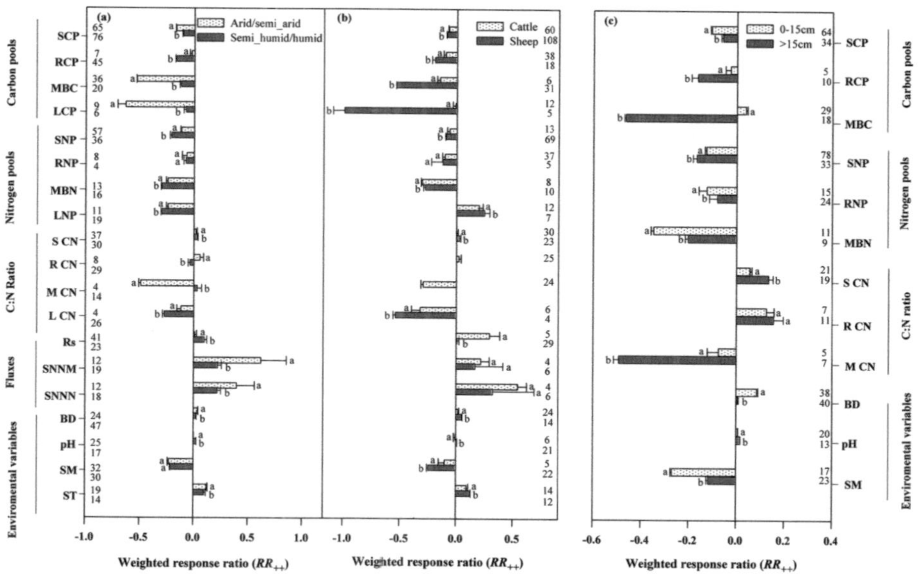

Figure 5.
*Weighted response ratio (RR_{++}) of 19 variables related to carbon and nitrogen cycles in response to arid/semi-arid-white columns and semi-humid/humid-gray columns (**a**), cattle grazing—white columns and sheep grazing—gray columns (**b**), 0-15 cm - white columns and > 15 cm—gray columns (c). Bars represent RR_{++} ± 95% confidence intervals. The vertical line was drawn at RR_{++} = 0. Numbers for each bar indicate the sample size. Symbols* a *and* b *represent the significant differences among two categories (panel a, arid/semi-arid vs. semi-humid/humid climate; panel b, cattle vs. sheep grazing; panel c, soil depth of 0–15 cm vs. >15 cm) for the responses of selected variables to grazing. SCP, soil carbon pools; RCP, root carbon pools; MBC, microbial carbon; LCP, litter carbon pools; SNP, soil nitrogen pools; MBN, microbial nitrogen; LNP, litter nitrogen pools; RNP, root nitrogen pools; SCN, soil C:N ratio; MCN, microbial C:N ratio; RCN, root C:N ratio; Rs, soil respiration; SNNM, soil net mineralization; SNNN, soil net N nitrification; BD, bulk density; SM, soil moisture; ST, soil temperature.*

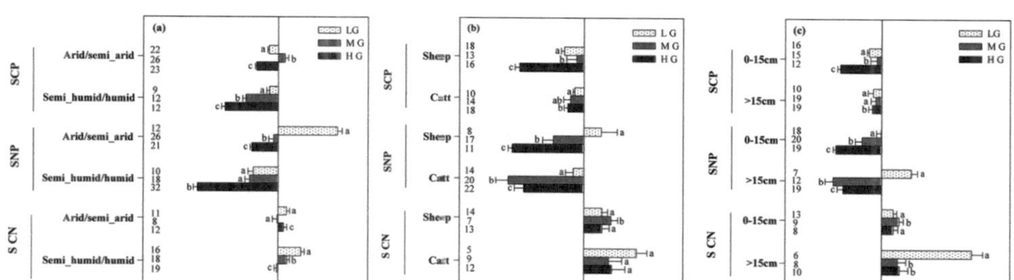

Figure 6.
Weighted response ratio (RR_{++}) of soil carbon pools (SCP), soil nitrogen pools (SNP), and soil C:N ratio (SCN) in different grazing intensities with respect to climate type (a), livestock type (b) and soil depth (c). Bars represent RR_{++} ± 95% confidence intervals. The vertical line was drawn at RR_{++} = 0. Number values for each bar indicate the sample size. Symbols a, b *and* c *represents the significant differences among three grazing intensities for the responses of selected variables to grazing.*

Livestock type, climate type, and soil depth also affected the overall magnitude and even direction of the weighted response ratios of SCP, SNP as well as SCN under different grazing intensities (**Figure 6**). The meta-analysis shows that both SCP and SNP in semi-humid/humid regions decreased with increasing intensity, whereas moderate and light grazing exhibited positive effects on SCP and SNP in arid/semi-arid regions. Decreased SCP was highest under heavy grazing, followed by light and moderate grazing, irrespective of cattle or sheep grazing. Light grazing

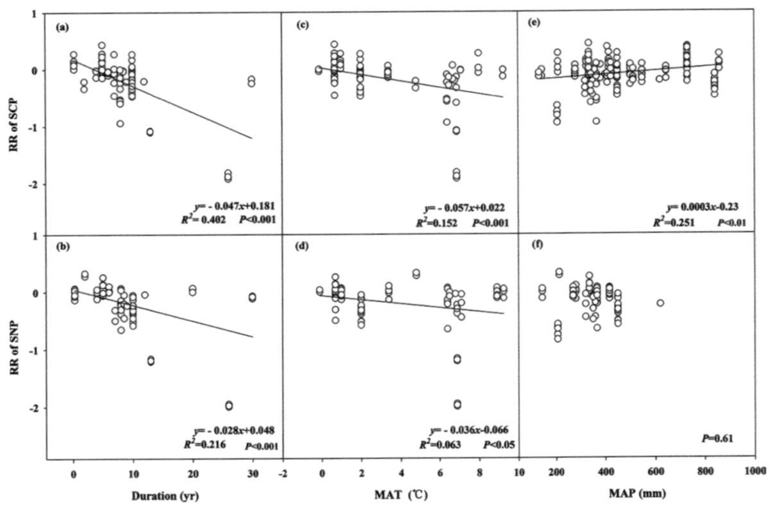

Figure 7.
Relationships of grazing duration (a, b), mean annual temperature (MAT, c, d), and mean annual precipitation (MAP, e, f) with response ratios (RR) of soil carbon pools (SCP, a, c, e) and soil nitrogen pools (SNP, b, d, f).

exhibited positive effects on SNP at the depth of >15 cm, while both moderate and heavy grazing had the opposite effects on it at the same depth (**Figure 6**). These differences induced by livestock type, climate type and soil depth may results from the complex interaction between grazing intensity with water, temperature and nutrients, but the potential mechanisms was still unknown and need further investigations.

4. Implication for grassland management

Overgrazing is a primary contributor to grassland degradation and desertification, which may significantly affect ecosystems functions and then lead to positive or negative climate-biosphere feedbacks [8, 25]. The regional and global studies showed that grazing intensity is a very important role in regulating belowground C and N pools and fluxes, which may offer some suggestions for future grassland management and model development. First, the effects of grazing intensity on C and N cycles may be regulated by environmental conditions (e.g., nitrogen and water availability; [8]). However, how the interactions of grazing with global change factors (e.g., warming, nitrogen addition, elevated CO_2, increased precipitation and drought) is influenced by grazing intensity remain unknown [44, 45]. These knowledge gaps may impede us to fully understand how grazing affects C and N cycles of grasslands at global scale.

Second, current global synthesized studies showed that most of current grazing studies were distributed in temperate climates, such as eastern Asia and North America, and only few studies were conducted in cold and tropical regions [5, 6]. Thus, more studies from other regions (e.g., Africa and Australia) should be conducted in order to develop a more comprehensive understanding of how grazing intensity influence C and N cycling of global grasslands. Another problem is the experimental duration. Most of current grazing experiments were less than 10 years, due to the high costs and long time scale. The grazing effects on C and N cycle may vary with time [5]. Hence, there is a need to conduct studies over one decade to better understand the effects of grazing on belowground C and N cycling.

P values of the correlations	Pearson correlation coefficients							
	***RR*(SCP)**	***RR*(SNP)**	***RR*(BD)**	***RR*(SM)**	**Latitude**	**MAP**	**MAT**	**Duration**
RR(SCP)		0.911**	−0.529**	0.135	−0.228*	0.201*	−0.474**	−0.634**
RR(SNP)	<0.001		−0.527**	0.080	−0.029	−0.055	−0.359**	−0.465**
RR(BD)	<0.001	<0.001		−0.742**	−0.367**	0.364**	−0.242	0.415**
RR(SM)	0.493	0.702	<0.001		0.546**	−0.366*	0.234	−0.267
Latitude	0.021	0.783	0.003	<0.001		−0.455**	0.396**	0.538**
MAP	0.045	0.610	0.005	0.033	<0.001		−0.656**	−0.061
MAT	<0.001	<0.001	0.052	0.190	<0.001	<0.001		0.359**
Duration	<0.001	<0.001	0.002	0.170	<0.001	0.571	<0.001	

The environmental/forcing variables are latitude, longitude, Mean annual precipitation (MAP), mean annual temperature (MAT), RRBD (Bulk Density), RRSM (Soil Moisture) and grazing time (Duration). The values on up-right side of the diagonal are Pearson correlation coefficients. The values on low-left side of the diagonal are P values to indicate statistical significance of the correlation coefficients.
Note: $^{*}P < 0.05$; $^{**}P < 0.01$; $^{***}P < 0.001$.

Table 1.
Correlation analysis of environmental variables with each other and with response ratio of SCP [RR(SCP)] and SNP [RR(SNP)] of surface soil (<15 cm).

Third, grazing intensity (light, moderate, and heavy grazing) significantly affects belowground C and N cycling in grassland ecosystems. Meanwhile, different combinations of grazing and global change factors (e.g., warming, nitrogen addition) also have disparate effects on C and N cycle of grasslands [8]. However, current land-surface models did not usually differentiate the effects of grazing intensities as well as their combinations with global change factors, which may trigger great challenges for us to predict the C-climate feedbacks in the Anthropocene. Therefore, future land-surface models may need thus to fully consider these processes in order to develop more precise process-based mechanism for forecasting the feedback of grassland ecosystems to climate change.

Fourth, environmental factors (both MAP and MAT) may be crucial in evaluating the response of belowground C and N cycling to different driving factors, as the effects of grazing, global change factors, and their combinations on belowground C and N cycling may change with MAT and MAP transects [6, 14]. The global study also demonstrated that response ratios of soil carbon content and soil nitrogen content to grazing in warmer biomes was clearly higher than those in the low range (**Figure 7**). These results demonstrated the importance of decreasing grazing frequency and intensity in warmer regions than colder ones, which may help to increase soil C sequestration in ecological fragile areas.

Acknowledgements

This research was financially supported by the "Thousand Young Talents" Program in China, National Natural Science Foundation of China (Grant No. 31600352, 31370489) and "Outstanding doctoral dissertation cultivation plan of action of East China Normal University (Grant No. YB2016023). We would like to acknowledge the permission from Global Change Biology to allow us to transfer our previous part work into monograph chapter to the publics.

Conflict of interest

All authors declare no conflict of Interests.

Author details

Guiyao Zhou[1], Lingyan Zhou[1] and Xuhui Zhou[1,2]*

1 Center for Global Change and Ecological Forecasting, Tiantong National Field Station for Forest Ecosystem Research, Shanghai Key Lab for Urban Ecological Processes and Eco-Restoration, School of Ecological and Environmental Sciences, East China Normal University, Shanghai, China

2 Shanghai Institute of Pollution Control and Ecological Security, Shanghai, China

*Address all correspondence to: xhzhou@des.ecnu.edu.cn

References

[1] Hufkens K, Keenan TF, Flanagan LB, Scott RL, Bernacchi CJ, Joo E, et al. Productivity of North American grasslands is increased under future climate scenarios despite rising aridity. Nature Climate Change. 2016;**6**:710-716

[2] Follett RF, Reed DA. Soil carbon sequestration in grazing lands: Societal benefits and policy implications. Rangeland Ecology and Management. 2010;**63**:4-15

[3] Salvati L, Carlucci M. Towards sustainability in agro-forest systems? Grazing intensity, soil degradation and the socioeconomic profile of rural communities in Italy. Ecological Economics. 2015;**112**:1-13

[4] Bai YF, Wu JG, Clark CM, Pan QM, Zhang LX, Chen SP, et al. Grazing alters ecosystem functioning and C:N:P stoichiometry of grasslands along a regional precipitation gradient. Journal of Applied Ecology. 2012;**49**:1204-1215

[5] Zhou GY, Zhou XH, He YH, Shao JJ, Hu ZH, Liu RQ, et al. Grazing intensity significantly affects belowground carbon and nitrogen cycling in grassland ecosystems: A meta-analysis. Global Change Biology. 2017;**23**:1167-1179

[6] Zhou GY, Luo Q, Chen YJ, Hu JQ, He M, Gao J, et al. Interactive effects of grazing and global change factors on soil and ecosystem respiration in grassland ecosystems: A global synthesis. Journal of Applied Ecology. 2019;**56**:2007-2019

[7] Yan L, Zhou GS, Zhang F. Effects of different grazing intensities on grassland production in China: A meta-analysis. PLoS One. 2013;**8**:e81466

[8] Zhou GY, Luo Q, Chen Y, He M, Zhou LY, Frank D, et al. Effects of livestock grazing on grassland carbon storage and release override impacts associated with global climate change. Global Change Biology. 2019a;**25**:1119-1132

[9] Connell JH. Diversity in tropical rain forests and coral reefs. Science. 1978;**199**:1302-1310

[10] Gong JR, Wang YH, Liu M, Huang YM, Yan X, Zhang ZY, et al. Effects of land use on soil respiration in the temperate steppe of Inner Mongolia, China. Soil and Tillage Research. 2014;**144**:20-31

[11] Knops JMH, Tilman D. Dynamics of soil nitrogen and carbon accumulation for 61 years after agricultural abandonment. Ecology. 2000;**81**:88-98

[12] Liu N, Kan HM, Yang GW, Zhang YJ. Changes in plant, soil, and microbes in a typical steppe from simulated grazing: Explaining potential change in soil C. Ecological Monographs. 2015;**85**:269-286

[13] Bagchi S, Ritchie ME. Introduced grazers can restrict potential soil carbon sequestration through impacts on plant community composition. Ecology Letters. 2010;**13**:959-968

[14] Mcsherry ME, Ritchie ME. Effects of grazing on grassland soil carbon: A global review. Global Change Biology. 2013;**19**:1347-1357

[15] McNaughton S, Ruess R, Seagle S. Large mammals and process dynamics in African ecosystems. Bioscience. 1988;**38**:794-800

[16] Zhou XQ, Wang JZ, Hao YB, Wang YF. Intermediate grazing intensities by sheep increase soil bacterial diversities in an inner Mongolian steppe. Biology and Fertility of Soils. 2010;**46**:817-824

[17] Schuman GE, Reeder JD, Manley JT, Hart RH, Manley WA. Impact of grazing

management on the carbon and nitrogen balance of a mixed-grass rangeland. Ecological Applications. 1999;**9**:65-71

[18] Liu N, Zhang YJ, Chang SJ, Kan HM, Lin LJ. Impact of grazing on soil carbon and microbial biomass in typical steppe and desert steppe of Inner Mongolia. PLoS One. 2013;7:e36434

[19] Manley J, Schuman G, Reeder JD, Hart RH. Rangeland soil carbon and nitrogen responses to grazing. Journal of Soil and Conservation. 1995;**50**:294-298

[20] Stavi I, Ungar ED, Lavee H, Lavee H, Sarah P. Grazing-induced spatial variability of soil bulk density and content of moisture, organic carbon and calcium carbonate in a semi-arid rangeland. Catena. 2008;75:288-296

[21] Zhang T, Zhang YJ, Xu MJ, Zhu JT, Wimberly MC, Yu GR, et al. Light-intensity grazing improves alpine meadow productivity and adaption to climate change on the Tibetan plateau. Scientific Reports. 2015;**5**:15949

[22] Derner J, Briske D, Boutton T. Does grazing mediate soil carbon and nitrogen accumulation beneath C_4, perennial grasses along an environmental gradient? Plant and Soil. 1997;**191**:147-156

[23] Baker JM, Ochsner TE, Venterea RT, Griffis TJ. Tillage and soil carbon sequestration—What do we really know? Agriculture, Ecosystems and Environment. 2007;**118**:1-5

[24] Reicosky DC. Tillage-induced CO_2 emission from soil. Nutrient Cycling in Agroecosystems. 1997;**49**:273-285

[25] He NP, Zhang YH, Yu Q, Chen SP, Pan QM, Zhang GM, et al. Grazing intensity impacts soil carbon and nitrogen storage of continental steppe. Ecosphere. 2011;**2**:304-316

[26] Parton WJ, Schimel DS, Cole CV, Ojima DS. Analysis of factors controlling soil organic matter levels in Great Plains grasslands. Soil Science Society of America Journal. 1987;**51**:1173-1179

[27] Wu HH, Wiesmeier M, Yu Q, Steffens M, Han XG, Kögel-Knabner I. Labile organic C and N mineralization of soil aggregate size classes in semiarid grasslands as affected by grazing management. Biology and Fertility of Soils. 2012;**48**:305-313

[28] Detling J, Dyer M, Winn D. Net photosynthesis, root respiration, and regrowth of *Bouteloua gracilis* following simulated grazing. Oecologia. 1979;**41**:127-134

[29] Knops JMH, Bradley KL, Wedin DA. Mechanisms of plant species impacts on ecosystem nitrogen cycling. Ecology Letters. 2002;**5**:454-466

[30] Savadogo P, Sawadogo L, Tiveau D. Effects of grazing intensity and prescribed fire on soil physical and hydrological properties and pasture yield in the savanna woodlands of Burkina Faso. Agriculture Ecosystems and Environment. 2007;**118**:80-92

[31] Thomey ML, Collins SL, Vargas R, Johnson JE, Brown RF, Natvig DO, et al. Effect of precipitation variability on net primary production and soil respiration in a Chihuahuan Desert grassland. Global Change Biology. 2011;**17**:1505-1515

[32] Chapin IIFS, Matson PA, Mooney HA. Principles of Terrestrial Ecosystem Ecology. New York, NY, USA: Springer; 2002

[33] Su YZ, Li YL, Cui HY, Zhao WZ. Influences of continuous grazing and livestock exclusion on soil properties in a degraded sandy grassland, Inner Mongolia, northern China. Catena. 2005;**59**:267-278

[34] Osem Y, Perevolotsky A, Kigel J. Site productivity and plant size explain the response of annual species to grazing exclusion in a Mediterranean semi-arid rangeland. Journal of Ecology. 2004;**92**:297-309

[35] Shi XM, Li XG, Li CT, Zhao Y, Shang ZH, Ma QF. Grazing exclusion decreases soil organic C storage at an alpine grassland of the Qinghai–Tibetan plateau. Ecological Engineering. 2013;**57**:183-187

[36] Xia JY, Niu SL, Wan SQ. Response of ecosystem carbon exchange to warming and nitrogen addition during two hydrologically contrasting growing seasons in a temperate steppe. Global Change Biology. 2009;**15**:1544-1556

[37] Zhou XH, Talley M, Luo YQ. Biomass, litter, and soil respiration along a precipitation gradient in southern Great Plains, USA. Ecosystems. 2009;**12**:1369-1380

[38] Hu ZM, Li SG, Guo Q, Niu SL, He NP, Li LH, et al. A synthesis of the effect of grazing exclusion on carbon dynamics in grasslands in China. Global Change Biology. 2016;**22**:1385-1393

[39] Luyssaert S, Inglima I, Jung M. CO_2 balance of boreal, temperate, and tropical forests derived from a global database. Global Change Biology. 2007;**13**:2509-2537

[40] Williams M, Eugster W, Rastetter EB, Mcfadden JP, Chapin FS III. The controls on net ecosystem productivity along an Arctic transect: A model comparison with flux measurements. Global Change Biology. 2000;**6**:116-126

[41] Wang CK, Yang JY, Zhang QZ. Soil respiration in six temperate forests in China. Global Change Biology. 2006;**12**:2103-2114

[42] Holland EA, Detling JK. Plant response to herbivory and belowground nitrogen cycling. Ecology. 1990;**71**:1040-1049

[43] Li CL, Hao XY, Zhao ML, Han GD, Willms WD. Influence of historic sheep grazing on vegetation and soil properties of a desert steppe in Inner Mongolia. Agriculture Ecosystems and Environment. 2008;**128**:109-116

[44] Lu M, Zhou XH, Luo YQ, Yang YH, Fang CM, Chen JK, et al. Minor stimulation of soil carbon storage by nitrogen addition: A meta-analysis. Agriculture Ecosystems and Environment. 2011;**140**:234-244

[45] Zhou LY, Zhou XH, Zhang BC, Lu M, Luo YQ, Liu LL, et al. Different responses of soil respiration and its components to nitrogen addition among biomes: A meta-analysis. Global Change Biology. 2014;**20**:2332-2343

Chapter 6

Small Grains as Winter Pasture in the Southern Great Plains of the United States

Tadele T. Kumssa, Joshua D. Anderson, Twain J. Butler and Xue-Feng Ma

Abstract

Small-grain cereals are widely adapted and used as annual cool-season pastures in the Southern Great Plains (SGP) of the United States, where livestock and forage production are the largest contributors to agricultural income. The advantage of growing small grains in the region is evident due to the widespread adoption and flexibility of production for grain only, forage only, or both grain and forage (i.e., dual purpose). Farmers in the SGP often prefer the use of small grains for dual purpose mainly because of alternative income options from livestock and/or grain, ensuring stable income especially when product prices fluctuate with market demands. Small-grain forage is exceptionally important during autumn, winter, and early spring when forage availability from other sources is low. By providing nutritionally high-quality forage, small grains minimize the need for protein and energy supplements. Besides being used for winter pasture, small grains also serve as cool-season cover crops. While small grains offer different advantages in the integrated crop-livestock system in the region, farming management practices can play an important role to maximize the benefit. The objectives of this chapter are to summarize the significance of small grains as winter pasture and highlight the production status of each small-grain species in the SGP of the United States.

Keywords: forage, oat, rye, small grains, Southern Great Plains, triticale, United States, wheat

1. Introduction

Small grains, such as wheat (*Triticum aestivum* L.), rye (*Secale cereale* L.), triticale (*X Triticosecale* Wittmack), and oat (*Avena sativa* L.), are an integral part of the forage-livestock system in the Southern Great Plains (SGP) of the United States, as they can be grazed during cool-season months when other forage species are not productive. On average in the last 3 years (2016–2018), 7 million hectares of land was planted annually by wheat alone for forage and grain production in the SGP, including Kansas, Oklahoma, and Texas [1], which is the largest area of low-rainfall winter wheat cropland worldwide. The SGP (32–40°N; 96–104°W) is generally classified as grassland, cropland, and forest land [2]. Although many crop species grow in the area, winter wheat covers the largest amount of cropland in the region. Small grains are well adapted to the SGP's environment, for both forage and grain (i.e., dual purpose) production. The warm autumn and mild winter with little snow coverage enable the small grains to grow fast and serve as ideal cool-season grazing pasture.

As a dual-purpose crop, small-grain cereals have other unnoticed advantages, such as weed and disease suppression by grazing animals. However, grazing small grains at the vegetative stage can reduce grain yield later in the summer depending on management practices and growing conditions, especially moisture. Farmers in the SGP often prefer the use of small grains for dual purpose mainly because of alternative income options from livestock and/or grain, ensuring relatively stable income especially when product prices fluctuate with market demands. As one of the crop-livestock integration methods, dual-purpose cereal crops can advance sustainable development through increased environmental balance, livelihood diversification, and flexibility to economic stresses, thus reducing risks [3].

Another aspect of crop-livestock integration is the use of cereal as both a cover crop and forage where plants seasonally cover the soil while serving as forage. Cover crops protect the soil primarily from wind and water erosion, improve soil quality, and increase nutrient cycling [4–6]. While small grains offer different advantages in the integrated crop-livestock system in the SGP, farming management practices, such as planting date and grazing intensity and duration, which depend on the weather and growth stage of the crop, can maximize the benefits [7, 8]. However, understanding overall management practices and fitting crops to climatic conditions is necessary for reliable accomplishment [9]. In this chapter, we present the significance of small grains as winter pasture in the SGP.

2. Agroecology and farming systems in the Southern Great Plains

The SGP of the United States extends from east of the Rocky Mountains in Colorado and New Mexico to Oklahoma and Kansas. The region also includes the Texas Panhandle and adjoining areas of West Texas and eastern New Mexico [10, 11]. The majority of the region is represented by Kansas, Oklahoma, and Texas. Agriculture is the most important land use in the region, where farming is more common in the east, while ranching dominates in the western part. The region mainly consists of mixed-grass and shortgrass terrestrial ecosystems. The main agricultural activity in the shortgrass prairie is animal grazing, while wheat cultivation dominates in the mixed-grass prairie [12]. Overtime, much of the native ecosystems have been converted to either farmland or pasture grazed by domestic livestock [13]. In 2012, the value of agricultural activities in the region surpassed $59 billion, of which livestock contributed more than half of the total [2]. Farming is generally water-limited in the region, as drought occasionally occurs and impacts the whole agricultural system [14, 15].

The climate of the SGP is typically characterized with low precipitation, high evaporation rate, windy hot summers, and a wide range of daily and seasonal temperatures [10, 11]. The weather is highly variable, spanning from extreme cold to extreme heat and from very limited or no rain to extreme rainfall or flooding conditions, which often affect the overall socioeconomic system in the region [16]. These extreme weather conditions sometimes enable proliferation of invasive weeds and pests, leading to ecosystem imbalance.

Although temperature and precipitation are highly variable, making it difficult to define a crop that performs well region-wide [17], the mild winter favors cool-season crops for grazing animals [18]. The region is known for its mixed farming practices, producing both livestock and grain. The most common crop-livestock integration methods in the United States are sod-based crop rotations, livestock grazing of crops within cash crop rotations, grazing of crop residues, sod intercropping, dual-purpose cereal crops, and silvopasture [19]. Wherever it is practiced, crop-livestock integration in the United States can be implemented either within the farm, in which crops and livestock produced at the same time and space, or among the farms, in which farmers work together in agreements to attain the intended synergies between production systems [20]. Dual-purpose cereal production has been the preferred system in the SGP because it gives relatively stable income [21]. In dual-purpose settings, small grains are grazed without removing terminal meristems so that plants regenerate and grow for grain production. Based on grain and cattle (*Bos* spp.) prices, producers decide whether to use small grains as forage only, grain only, or both forage and grain, enabling flexibility to ensure greater income.

3. Significance of winter small grains as pasture in the Southern Great Plains

Small-grain forage is important in agricultural systems worldwide. In some parts of the world, intensive grazing of small-grain pasture occurs only once but continuously for days [22–24]. In the Central and Southern Great Plains of the United States, however, grazing is often at lower stocking rates for an extended period of time (months) [25, 26]. The grassland in the SGP is predominantly warm-season species, and hence the cool-season small-grain pasture plays critical roles in supplying nutritionally high-quality forage in winter when the quantity and quality of warm-season forage become low [27, 28].

Forage-based livestock production is an important element of the SGP agricultural economy [29]. This region plays an important role in the beef industry of the United States due to the strategic location between the humid southeast where a majority of the cow/calf enterprises reside and the High Plains region where the grain production and the finishing feedlots exclusively reside. Millions of stocker calves from more than 500,000 farms across the southern United States pass through the SGP on their way to the feedlots [30]. Managing small grains for both grazing and grain is an alternative and sustainable practice that supports the crop-livestock system in the region [30–32]. Cool-season small-grain forage provides more flexible and profitable crop-livestock systems, in which producers can adjust production outcomes they want based on market values of grains and livestock [28, 33]. In the past, it has been estimated that 30 to 80% of the total wheat area planted in the SGP is grazed at some time during the growing season and 10 to 20% of the area is grazed throughout the spring and not harvested for grain, which is referred to as "graze-out" [34]. When combined with summer annual forages, the winter annual small-grain forages can result in the best net return in unit area of land [35].

Small grains generally grow faster in winter than most other pastures and can also recover after grazing [36]. Small grains are also known to have high tillering capacity that enables them to tolerate stress from grazing. The average autumn-winter forage yield of small grains in the SGP is about 2500–3500 kg ha^{-1}; however, large yield variation was observed depending on growing conditions [37, 38]. In addition, small-grain forage is high in protein content and digestibility and hence promotes more weight gain in grazing animals [39]. In Texas, Oklahoma, and Kansas, small grains are a good source of high-quality pasture from late autumn to early spring [40]. Several studies indicated that small grains produce about 1.0 kg d^{-1} of average daily gain (ADG) in grazing animals [41, 42].

To maximize the advantage of small grains as winter pasture, proper management practices should be followed. For example, early planting in a prepared seedbed is important to ensure early availability of the pasture for grazing. However, early planting may not be feasible if there are biotic and abiotic stresses in the farm.

Grazing too early when plants are small (not well-rooted) leads to pulling up of plants and also severe damage from trampling. In the dual-purpose production system, termination of grazing at the first hollow stem (jointing) stage is critical to achieve the best economic return [43]. Delaying the termination of grazing will reduce subsequent grain yield by approximately as much as five percent per day for every day past the first hollow stem [8].

Although winter wheat is the main cool-season forage source in the region, other small grains in general are also important for their wide adaptability and versatility in forage use, such as pasture, green cut, silage, and hay. The most commonly grown small grains for winter pasture in the region are wheat, rye, triticale, and oat. Compared to wheat, other small grains (rye, triticale, and oat) can produce greater forage yield in a specific season (autumn, winter, or spring), and they have competitive advantage in graze-out systems. Farmers select the crop that fits the need of winter pasture based on the characteristics of the crops, available management practices, and production goals.

4. Small grains commonly grown in the Southern Great Plains

4.1 Wheat

Wheat is the largest and most important crop in the SGP, with 3.1, 1.8, and 1.8 million hectares being planted in Kansas, Oklahoma, and Texas, respectively, in 2018 [1]. Wherever growing conditions are favorable, the economic advantage of managing wheat as a dual-purpose crop is better than managing it as a grain-only or forage-only crop [44, 45]. Wheat cultivars, such as "Endurance" and "Duster," with superior dual-purpose production have been released by the Oklahoma Agricultural Experiment Station [46, 47].

Obviously, the main reason why wheat is the most commonly planted small-grain species for grazing is its dual-purpose economic flexibility in the SGP. However, wheat is below average in amount of early growth and total forage yield compared to other small grains. Wheat is considered more tolerant to wet clay soils than the other small-grain species. Wheat forage is high in protein, energy, and minerals but low in fiber. To maximize forage yield in a dual-purpose winter wheat production system, the crop is often planted early in the autumn so that grazing can begin in late autumn. It is also very important to stop grazing at the first hollow stem stage when dormancy of the crop is released. Depending on the availability of moisture, wheat pasture can last 4–5 months.

4.2 Rye

Although rye acreage in the United States is much lower than wheat and oat acreages, it is an important forage crop in the SGP. Only 20% of the rye acreage in the United States is harvested [1], indicating that the majority of the crop is used for grazing animals. Oklahoma is the largest rye-growing state in the United States [1], at least partially because of the historical contribution of the Noble Research Institute in releasing a few well-known rye cultivars, such as "Elbon" (1956), which is still widely grown today. The Noble Research Institute has also released "Maton" (1975), "Oklon" (1993), "Bates" (1995), "Maton II" (2007), and "Bates RS4" (2013) rye cultivars, which were primarily selected for increasing autumn-winter production.

Among all small-grain cereals, rye is the most winter-hardy crop. It is proved to be the best in performance, especially under stressed growing conditions, because of its excellent biotic stress tolerance to multiple diseases and abiotic stress tolerance to frost, drought, low pH, and marginal soil fertility [48, 49]. Rye performs better than other small grains, especially in the light-textured sandy soils, due to its prolific root system. It grows faster and produces more forage than wheat from autumn to winter. Livestock producers consider rye the most dependable cereal for winter grazing because of its reliable and great forage production [50, 51] and lower potential for causing grass tetany (livestock disease caused by magnesium deficiency) [52].

The main drawback of rye is that it is considered a terrible weed in wheat grown for grain, so it should not be planted in areas expected to be used for wheat grain production in the future. In addition, since rye develops fast and matures early, rye flowers often encounter late-spring freezing damage, causing potential seed yield loss.

The progress of increasing rye forage yield has been minimal because of limited breeding efforts, as only a few institutes, such as the University of Florida and the Noble Research Institute, have rye breeding programs. Cultivars commonly grown in the southern United States are those released many years ago. Rye is also an important grain crop worldwide, with 75% of grain production from Russia, Poland, Germany, and Ukraine [48].

In addition, rye is the most widely grown cover crop for sustainable agriculture because of its competitive ability to suppress weeds and scavenge residual soil nitrogen after other crops. Cereal rye is used widely as a winter annual cover crop in the United States because of its winterhardiness, high biomass production, and residues against weed species [53]. Studies have indicated that rye cover crop improves soil organic matter, nitrogen mineralization, and particulate organic matter [54].

4.3 Triticale

Triticale is a man-made crop from hybridizing wheat (*Triticum* spp. L.) and rye for combining the best traits from the two parental species. This artificial cereal inherited its cold tolerance, disease resistance or tolerance, and adaptation to unfavorable soil and climatic conditions from the male parent, rye, and its yield and nutritional quality from the female parent, wheat [55, 56]. Although breeding efforts on triticale have been limited compared to those on other small grains, improved triticale cultivars produce competitive biomass and grain yield, making it a viable alternative crop especially under unfavorable growing conditions with diverse biotic and abiotic stress factors [56].

Triticale is an important crop especially in livestock farming systems, in which the crop is mainly produced for animal feed grain and/or forage [57–60]. In the SGP, triticale is grown mainly for grazing because of its superior performance in forage biomass production [56]. The forage yield of triticale is similar to that of rye but greater than that of wheat and oat in the southern Oklahoma [61]. Triticale also has greater forage quality for grazing animals when compared to rye [56]. In addition, triticale is a preferred forage crop for producing silage to cover the forage quality gap during dry, hot summers because of its high yielding and nutritional forage [56, 62]. Therefore, triticale is becoming more popular as an alternative forage to wheat and rye in the SGP, and it has the greatest potential for improvement due to its short breeding history [56].

4.4 Oats

Oat is another cereal commonly grown for grain and forage throughout the world. It is produced mainly in temperate, cool, and subtropical climates for grain and different forms of animal feed [63]. In the United States, spring oat is primarily produced for grain, while winter type is often grown for forage and in some cases as a dual-purpose crop [63].

The use of winter oat as pasture or forage is common in the SGP of the United States [64]. Oat is an important forage crop in the region and is a useful alternative to forage wheat for stocker cattle production [63]. Oat grows fast and generates very competitive forage biomass yield in autumn and spring when air temperature is optimal for the crop; however, it produces much less forage in winter than wheat, rye, and triticale because of its sensitivity to freezing temperatures. Therefore, improving winterhardiness is one of the main breeding objectives for winter forage oat. In addition, compared to other small grains, oat is in general more susceptible to various diseases, and it does not grow well in low-input systems; thus, it is not recommended for marginal land. Oat is generally the most preferred forage to grazing animals among all small grains due to its superior palatability.

In summary, pasture production varies greatly among small-grain cereals. Comparing forage, especially for stocker development, requires different

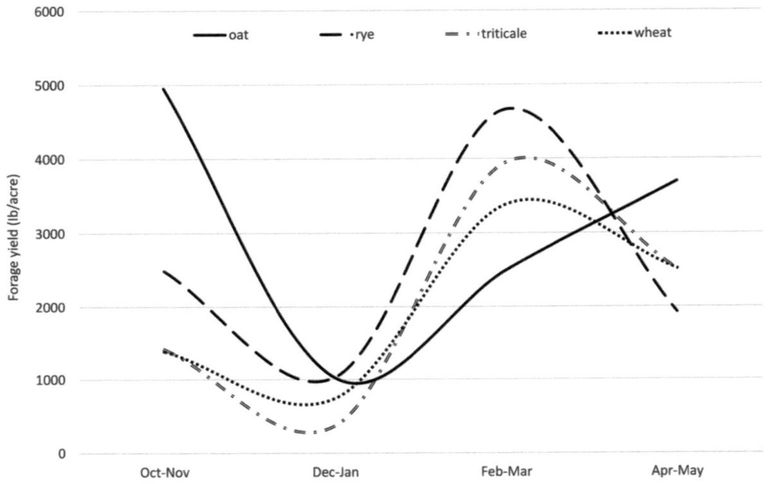

Figure 1.
Relative pasture production distribution of cool season small grain crops on heavy soil at Ardmore, Oklahoma.

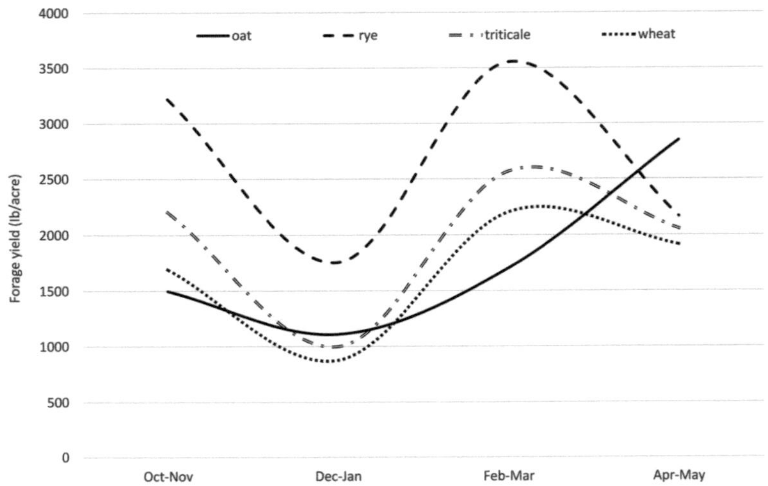

Figure 2.
Relative pasture production distribution of cool season small grain crops on sandy soil at Burneyville, Oklahoma.

considerations specific to the farm, growing season, and climatic conditions. For example, oat is the best for autumn forage yield on clay soil at Ardmore, Oklahoma, while rye is the best on sandy soil at Burneyville, Oklahoma (**Figures 1** and **2**). Therefore, a holistic understanding of production environment, available management options, and economic analysis is key to attain the intended use. The availability of alternative forage options of small grains enables producers to make decisions that best fit their specific farm and climatic conditions.

5. Improving small grains for winter pasture in the Southern Great Plains

Since small-grain cereals are produced mainly for grain, breeding efforts to improve small grains mostly focus on increasing grain yield and quality worldwide. Although small-grain cereals are also widely used as forage crops, cultivars being used for winter pastures were mostly developed for grain rather than forage. Therefore, there is a need to improve small grains for winter forage or dual-purpose production, especially in the SGP where livestock and forage production are the largest contributors to agricultural income. Ideally, small grains for grazing are those with tolerance to early planting, animal grazing, and various seedling biotic and abiotic stresses and those with vigorous early growth and regrowth and/or extended periods of vegetative growth.

In the southern United States, institutes that have breeding programs to improve cereals for forage or dual purpose include the Noble Research Institute (formerly the Samuel Roberts Noble Foundation), the Oklahoma State University, the Texas A&M AgriLife Research (Texas A&M University), the University of Georgia, and the University of Florida [46, 63, 65]. However, only the Noble Research Institute has a forage-focused small-grain breeding program, while other universities mainly deliver dual-purpose cultivars from their grain-focused breeding programs. The Noble Research Institute started its small-grain breeding program with rye in the early 1950s. Since then, the institute has expanded its breeding program to include wheat, triticale, and oat with the objective of increasing forage or dual-purpose production. The research has particularly focused on developing cultivars with improved early vigor and regrowth vigor, improved grazing tolerance, and increased autumn-winter forage yield or total forage yield. The program has developed multiple improved forage cultivars of wheat, rye, oat, and triticale.

Biomass yield is the main target for forage crop improvement. The trait is usually measured through multiple clipping over seasons, posing difficulty in biomass yield estimation, especially when a large number of samples need to be evaluated at field scales [63]. Other physical methods of estimating biomass involved rising plate meters, capacitance meters, and clipping samples with meter sticks [66–69]. However, not only are these methods time-consuming, but it is also difficult to establish a reliable model for estimating biomass yield [70].

To augment breeding efficiency, high-throughput phenotyping platforms using remote sensing have been adopted recently in forage breeding and have facilitated biomass yield estimation for breeding selection [70–73]. Over the last few years, the Noble Research Institute has developed ground-based high-throughput phenotyping platforms and improved biomass prediction accuracies by incorporating normalized difference vegetation index (NDVI) with proximal sensors, such as ultrasonic and laser height measurements [70]. Similar phenotyping platforms have also been used in other breeding programs.

6. Summary

Cool-season annual small grains remain the most important forage for stocker cattle producers in the SGP of the United States due to their high forage quality, adaptation to the environment, and economic contribution. They are also valuable for filling forage gaps and extending the grazing season as a complement to other grasses. Wheat, rye, triticale, and oat are the most common small-grain forages that provide production flexibility and economic stability with an alternative dual-purpose option. Wheat is the most commonly grown because of its alternative value as a cash crop; rye is currently the best forage option on sandy, drought-prone soils or in graze-out systems; oat can be the most valuable when providing autumn forage; and triticale has the greatest potential for improvement and increased adoption.

Acknowledgements

The authors sincerely thank Andrea Mongler for critical reading of the manuscript.

Conflict of interest

The authors declare no conflicts of interest.

Author details

Tadele T. Kumssa, Joshua D. Anderson, Twain J. Butler and Xue-Feng Ma*
Noble Research Institute, Ardmore, OK, USA

*Address all correspondence to: xma@noble.org

References

[1] USDA-NASS. United States Department of Agriculture Data and Statistics. 2019. Available from: https://www.nass.usda.gov/Quick_Stats/Lite/

[2] Steiner JL, Briske DD, Brown DP, Rottler CM. Vulnerability of Southern Plains agriculture to climate change. Climatic Change. 2017;**146**(1):201-218. DOI: 10.1007/s10584-017-1965-5

[3] Abson DJ, Fraser EDG, Benton TG. Landscape diversity and the resilience of agricultural returns: A portfolio analysis of land-use patterns and economic returns from lowland agriculture. Agriculture and Food Security. 2013;**2**(1):2. DOI: 10.1186/2048-7010-2-2

[4] Stevens CJ, Quinton JN. Diffuse pollution swapping in arable agricultural systems. Critical Reviews in Environmental Science and Technology. 2009;**39**(6):478-520. DOI: 10.1080/10643380801910017

[5] Edwards L. Comparison of two spring seeding methods to establish forage cover crops in relay with winter cereals. Soil and Tillage Research. 1998;**45**(3-4):227-235. DOI: 10.1016/S0167-1987(97)00065-2

[6] Blanco-Canqui H, Holman JD, Schlegel AJ, Tatarko J, Shaver TM. Replacing fallow with cover crops in a semiarid soil: Effects on soil properties. Soil Science Society of America Journal. 2013;**77**(3):1026-1034. DOI: 10.2136/sssaj2013.01.0006

[7] Edwards JT, Carver BF, Horn GW, Payton ME. Impact of dual-purpose management on wheat grain yield. Crop Science. 2011;**51**(5):2181-2185. DOI: 10.2135/cropsci2011.01.0043

[8] Edwards J, Horn G. First hollow stem: A critical wheat growth stage for dual-purpose producers. Crops and Soils. 2016;**49**(1):10-14. DOI: 10.2134/cs2016-49-1-2

[9] Franzluebbers A. Integrated crop–livestock systems in the southeastern USA. Agronomy Journal. 2007;**99**(2):361-372. DOI: 10.2134/agronj2006.0076

[10] Savage D, Costello D. The Southern Great Plains: The region and its need. In: Yearbook of Agriculture. Washington, DC: USDA; 1948. pp. 503-506

[11] Paysen TE, Ansley RJ, Brown JK, Gottfried GJ, Haase SM, Harrington MG, et al. Fire in western shrubland, woodland, and grassland ecosystems. In: Brown JK, Smith JK, editors. Wildland Fire in Ecosystems: Effects of Fire on Flora 2. U.S. Department of Agriculture, Forest Service, Rocky Mountain Research Station; Scotts Valley, CA, USA: CreateSpace publishing; 2000. pp. 121-159

[12] Hart RH. Land-use history on the short grass steppe. In: Ecology of the Short Grass Steppe: A Long-Term Perspective. New York: Oxford University Press; 2008

[13] Assal TJ, Melcher CP, Carr NB. Southern Great Plains Rapid Ecoregional Assessment: Pre-Assessment Report. US Geological Survey, 2015. 2331-1258

[14] Raz-Yaseef N, Billesbach DP, Fischer ML, Biraud SC, Gunter SA, Bradford JA, et al. Vulnerability of crops and native grasses to summer drying in the US Southern Great Plains. Agriculture, Ecosystems and Environment. 2015;**213**:209-218. DOI: 10.1016/j.agee.2015.07.021

[15] Christian J, Christian K, Basara JB. Drought and pluvial dipole events within the great plains of the United States. Journal of Applied

Meteorology and Climatology. 2015;**54**(9):1886-1898. DOI: 10.1175/JAMC-D-15-0002.1

[16] Kloesel K, Bartush B, Banner J, Brown D, Lemory J, Lin X, et al. In: Reidmiller DR, Avery CW, Easterling DR, Kunkel KE, Lewis KLM, Maycock TK, et al., editors. Southern Great Plains. Impacts, Risks, and Adaptation in the United States: Fourth National Climate Assessment, Volume II. Washington, DC, USA: U.S. Global Change Research Program; 2018. pp. 987-1035. DOI: 10.7930/NCA4.2018.CH23

[17] Baath GS, Northup BK, Rocateli AC, Gowda PH, Neel JP. Forage potential of summer annual grain legumes in the Southern Great Plains. Agronomy Journal. 2018;**110**(6):1-13. DOI: 10.2134/agronj2017.12.0726

[18] Mullenix M, Rouquette F. Cool-season annual grasses or grass–clover management options for extending the fall–winter–early spring grazing season for beef cattle. The Professional Animal Scientist. 2018;**34**(3):231-239. DOI: 10.15232/pas.2017-01714

[19] Sulc RM, Franzluebbers AJ. Exploring integrated crop–livestock systems in different ecoregions of the United States. European Journal of Agronomy. 2014;**57**:21-30. DOI: 10.1016/j.eja.2013.10.007

[20] Russelle MP, Entz MH, Franzluebbers AJ. Reconsidering integrated crop–livestock systems in North America. Agronomy Journal. 2007;**99**(2):325-334. DOI: 10.2134/agronj2006.0139

[21] Redmon LA, Horn GW, Krenzer EG, Bernardo DJ. A review of livestock grazing and wheat grain yield: Boom or bust? Agronomy Journal. 1995;**87**(2):137-147. DOI: 10.2134/agronj1995.00021962008700020001x

[22] Kelman W, Dove H. Growth and phenology of winter wheat and oats in a dual-purpose management system. Crop & Pasture Science. 2009;**60**(10):921-932. DOI: 10.1071/CP09029

[23] Francia E, Pecchioni N, Nicosia OLD, Paoletta G, Taibi L, Franco V, et al. Dual-purpose barley and oat in a Mediterranean environment. Field Crops Research. 2006;**99**(2-3):158-166. DOI: 10.1016/j.fcr.2006.04.006

[24] Anderson W. Production of green feed and grain from grazed barley in Northern Syria. Field Crops Research. 1985;**10**:57-75. DOI: 10.1016/0378-4290(85)90006-1

[25] Khalil IH, Carver BF, Krenzer EG, MacKown CT, Horn GW. Genetic trends in winter wheat yield and test weight under dual-purpose and grain-only management systems. Crop Science. 2002;**42**(3):710-715. DOI: 10.2135/cropsci2002.0710

[26] Holman JD, Thompson CR, Hale RL, Schlegel AJ. Grazing effects on yield and quality of hard red and hard white winter wheat. Agronomy Journal. 2009;**101**(4):775-788. DOI: 10.2134/agronj2008.0163x

[27] Moore AD, Bell LW, Revell DK. Feed gaps in mixed-farming systems: Insights from the grain & graze program. Animal Production Science. 2009;**49**(10):736-748. DOI: 10.1071/AN09010

[28] Ates S, Cicek H, Gultekin I, Yigezu Y, Keser M, Filley S. Bio-economic analysis of dual-purpose management of winter cereals in high and low input production systems. Field Crops Research. 2018;**227**:56-66. DOI: 10.1016/j.fcr.2018.08.003

[29] Rao S, Phillips W, Mayeux H, Phatak S. Potential grain and forage production of early maturing pigeonpea in the Southern Great Plains. Crop

Science. 2003;**43**(6):2212-2217. DOI: 10.2135/cropsci2003.2212

[30] MacKown CT, Carver BF. Fall forage biomass and nitrogen composition of winter wheat populations selected from grain-only and dual-purpose environments. Crop Science. 2005;**45**(1):322-328. DOI: 10.2135/cropsci2005.0322

[31] Giunta F, Cabigliera A, Virdis A, Motzo R. Dual-purpose use affects phenology of triticale. Field Crops Research. 2015;**183**:111-116. DOI: 10.1016/j.fcr.2015.07.026

[32] Bonachela S, Orgaz F, Fereres E. Winter cereals grown for grain and for the dual purpose of forage plus grain II. Water use and water-use efficiency. Field Crops Research. 1995;**44**(1):13-24. DOI: 10.1016/0378-4290(95)00046-3

[33] Garrett R, Niles M, Gil J, Gaudin A, Chaplin-Kramer R, Assmann A, et al. Social and ecological analysis of commercial integrated crop livestock systems: Current knowledge and remaining uncertainty. Agricultural Systems. 2017;**155**:136-146. DOI: 10.1016/j.agsy.2017.05.003

[34] Pinchak W, Worrall W, Caldwell S, Hunt L, Worrall N, Conoly M. Interrelationships of forage and steer growth dynamics on wheat pasture. Journal of Range Management. 1996;**49**(2): 126-130. DOI: 10.2307/4002681

[35] Braunwart K, Putnam D, Fohner G, editors. Alternative annual forages–now and in the future. In: Proc 31st California Alfalfa and Forage Symposium. 2001

[36] Hennessy G, Clements B. Cereals for grazing. NSW DPI: Prime Facts, 720; 2009

[37] Marburger D, Calhoun R, Pugh B, Watson B, Gillespie C. Fall Forage Production and First Hollow Stem Date in Small Grain Varieties during the 2017-2018 Crop Year. Oklahoma State University, Stillwater, OK, USA: Oklahoma Cooperative Extension Service; 2018. CR-2141

[38] Neely C, Hathcoat D, Gerrish B, Kimura E, Ramirez J, Berry M, et al. Forage variety results: Texas cool-season annual variety trials. Texas A&M University, TX, USA: Texas A&M AgriLife; 2018. SCS-2018-07

[39] Dove H, McMullen K. Diet selection, herbage intake and liveweight gain in young sheep grazing dual-purpose wheats and sheep responses to mineral supplements. Animal Production Science. 2009;**49**(10): 749-758. DOI: 10.1071/AN09009

[40] Hossain I, Epplin FM, Krenzer EG. Planting date influence on dual-purpose winter wheat forage yield, grain yield, and test weight. Agronomy Journal. 2003;**95**(5):1179-1188. DOI: 10.2134/agronj2003.1179

[41] Islam MA, Biermacher JT, Interrante SM, Reuter RR, Hopkins AA, Cook BJ, et al. Production and economics of grazing rye–annual ryegrass and tall fescue systems. Agronomy Journal. 2011;**103**:558-564. DOI: 10.2134/agronj2010.0325

[42] Beck PA, Hubbell DS, Watkins KB, Gunter SA, Daniels LB. Performance of stocker cattle grazing cool-season annual grass mixtures in northern Arkansas. The Professional Animal Scientist. 2005;**21**(6):465-473. DOI: 10.15232/S1080-7446(15)31251-1

[43] DeVuyst EA, Epplin FM, Taylor KW, Horn G, Edwards JT. Effect of Grazing Past First Hollow Stem on Wheat and Stocker Profits. Oklahoma State University, Stillwater, OK, USA: Oklahoma Cooperative Extension; 2011. AGEC-265

[44] Decker JE, Epplin FM, Morley DL, Peeper TF. Economics of five wheat production systems with no-till and conventional tillage. Agronomy Journal. 2009;**101**(2):364-372. DOI: 10.2134/agronj2008.0159

[45] Arzadun MJ, Arroquy JI, Laborde HE, Brevedan RE. Grazing pressure on beef and grain production of dual-purpose wheat in Argentina. Agronomy Journal. 2003;**95**(5):1157-1162. DOI: 10.2134/agronj2003.1157

[46] Edwards JT, Hunger RM, Smith EL, Horn GW, Chen M-S, Yan L, et al. 'Duster' wheat: A durable, dual-purpose cultivar adapted to the Southern Great Plains of the USA. Journal of Plant Registrations. 2012;**6**(1):37-48. DOI: 10.3198/jpr2011.04.0195crc

[47] Carver BF, Smith EL, Hunger RM, Klatt AR, Edwards JT, Porter DR, et al. Registration of 'Endurance' wheat. Crop Science. 2006;**46**(4):1816-1818. DOI: 10.2135/cropsci2005.12-0452CRC

[48] Geiger H, Miedaner T. Rye breeding. In: Carena MJ, editor. Cereals. 3. New York: Springer; 2009. pp. 157-181

[49] Bushuk W. Rye production and uses worldwide. Cereal Foods World. 2001;**46**(2):70-73

[50] Poysa V. Effect of forage harvest on grain yield and agronomic performance of winter triticale, wheat and rye. Canadian Journal of Plant Science. 1985;**65**(4):879-888. DOI: 10.4141/cjps85-113

[51] Blount A, Barnett R, Pfahler P, Johnson J, Buntin G, Cunfer B. Rye and Triticale Breeding in the South. Gainesville, FL: University of Florida; 2017

[52] Mayland H, Grunes D, Lazar V. Grass tetany hazard of cereal forages based upon chemical composition. Agronomy Journal. 1976;**68**(4):665-667. DOI: 10.2134/agronj1976.00021962006800040033x

[53] Mirsky SB, Curran WS, Mortensen DA, Ryan MR, Shumway DL. Control of cereal rye with a roller/crimper as influenced by cover crop phenology. Agronomy Journal. 2009;**101**(6):1589-1596. DOI: 10.2134/agronj2009.0130

[54] Moore E, Wiedenhoeft M, Kaspar T, Cambardella C. Rye cover crop effects on soil quality in no-till corn silage–soybean cropping systems. Soil Science Society of America Journal. 2014;**78**(3):968-976. DOI: 10.2136/sssaj2013.09.0401

[55] Walker AS, Bouguennec A, Confais J, Morgant G, Leroux P. Evidence of host-range expansion from new powdery mildew (*Blumeria graminis*) infections of triticale (× *Triticosecale*) in France. Plant Pathology. 2011;**60**(2):207-220. DOI: 10.1111/j.1365-3059.2010.02379.x

[56] Ayalew H, Kumssa TT, Butler TJ, Ma X-F. Triticale improvement for forage and cover crop uses in the Southern Great Plains of the United States. Frontiers in Plant Science. 2018;**9**:1130. DOI: 10.3389/fpls.2018.01130

[57] Salmon D, Mergoum M, Gomez-MacphersonH.Triticaleproduction and management. In: Mergoum M, Gomez-Maxpherson H, editors. Triticale Improvement and Production. Rome: FAO Plant Production and Protection Paper, Food and Agriculture Organization of the United Nations; 2004. pp. 27-36

[58] Saha MC, Baker JL, Bouton JH. Registration of 'NF201' forage triticale. Journal of Plant Registrations. 2015;**9**(2):185-189. DOI: 10.3198/jpr2014.10.0078crc

[59] Lekgari LA, Baenziger PS, Vogel KP, Baltensperger DD. Identifying

winter forage triticale (× *Triticosecale* Wittmack) strains for the central Great Plains. Crop Science. 2008;**48**(5):2040-2048

[60] Barnett R, Blount A, Pfahler P, Johnson J, Buntin G, Cunfer B. Rye and triticale breeding in the south. University of Florida, Gainesville, FL, USA: UF-IFAS Extension Service; 2002. SS-AGR-42

[61] Kim K-S, Webb S, Newell M, Anderson J, Butler T. Variation of winter forage production in four small grain species: Oat, rye, triticale and wheat. Pakistan Journal of Botany. 2017;**49**:553-559

[62] Delogu G, Faccini N, Faccioli P, Reggiani F, Lendini M, Berardo N, et al. Dry matter yield and quality evaluation at two phenological stages of forage triticale grown in the Po Valley and Sardinia, Italy. Field Crops Research. 2002;**74**(2):207-215. DOI: 10.1016/S0378-4290(02)00002-3

[63] Kim K-S, Tinker NA, Newell MA. Improvement of oat as a winter forage crop in the Southern United States. Crop Science. 2014;**54**(4):1336-1346. DOI: 10.2135/cropsci2013.07.0505

[64] Frasser J, McCartney D. Fodder oats in North America. In: Suttie JM, Reynolds SG, editors. Fodder Oats, a World Overview. Rome: Food and Agriculture Organization of the United Nations; 2002. pp. 19-36

[65] Newell MA, Butler TJ. Forage rye improvement in the southern United States: A review. Crop Science. 2013;**53**(1):38-47. DOI: 10.2135/cropsci2012.05.0319

[66] Dougherty M, Burger JA, Feldhake CM, AbdelGadir A. Calibration and use of plate meter regressions for pasture mass estimation in an Appalachian silvopasture. Archives of Agronomy and Soil Science. 2013;**59**(2):305-315. DOI: 10.1080/03650340.2011.615026

[67] Fehmi J, Stevens J. A plate meter inadequately estimated herbage mass in a semi-arid grassland. Grass and Forage Science. 2009;**64**(3):322-327. DOI: 10.1111/j.1365-2494.2009.00694.x

[68] Sanderson MA, Rotz CA, Fultz SW, Rayburn EB. Estimating forage mass with a commercial capacitance meter, rising plate meter, and pasture ruler. Agronomy Journal. 2001;**93**(6):1281-1286. DOI: 10.2134/agronj2001.1281

[69] Tucker CJ. A critical review of remote sensing and other methods for non-destructive estimation of standing crop biomass. Grass and Forage Science. 1980;**35**(3):177-182. DOI: 10.1111/j.1365-2494.1980.tb01509.x

[70] Pittman JJ, Arnall DB, Interrante SM, Moffet CA, Butler TJ. Estimation of biomass and canopy height in bermudagrass, alfalfa, and wheat using ultrasonic, laser, and spectral sensors. Sensors. 2015;**15**(2):2920-2943. DOI: 10.3390/s150202920

[71] Fricke T, Richter F, Wachendorf M. Assessment of forage mass from grassland swards by height measurement using an ultrasonic sensor. Computers and Electronics in Agriculture. 2011;**79**(2):142-152. DOI: 10.1016/j.compag.2011.09.005

[72] Fricke T, Wachendorf M. Combining ultrasonic sward height and spectral signatures to assess the biomass of legume–grass swards. Computers and Electronics in Agriculture. 2013;**99**:236-247. DOI: 10.1016/j.compag.2013.10.004

[73] Scotford I, Miller P. Combination of spectral reflectance and ultrasonic sensing to monitor the growth of winter wheat. Biosystems Engineering. 2004;**87**(1):27-38. DOI: 10.1016/j.biosystemseng.2003.09.009

Chapter 7

Performance of Forage Crops and Grass Mixtures in Kosovo

Sali Aliu, Imer Rusinovci, Shukri Fetahu and Dukagjin Zeka

Abstract

Grasslands represent a land-use which is effective and has great economical importance in the European agriculture. Grasslands these are important and effective sources of energy and proteins for ruminants, and combine high yields stability by draught resistance with low tillage operations and pesticide after use leading to good environmental conditions. Furthermore, good management practice in grasslands provides high potential of carbon sequestration in soils, resulting in climate change mitigation. Based on actual data, the meadows and pastures cover 224410.83 ha of total surface of the Kosovo; moreover there, just a particular of these surfaces are distinguished clearly as meadows or pastures. The most important forage crops which are dominant in our country include plants from *Poaceae*; this botanical family is divided into long, medium, and low herbs. Medium and long herbs are the best suited for mowing and silage or drying for sanitary preparations, while short herbs are very suitable for grassland. The most important of fodder or plants in Kosovo are: (*Medicago sativa* L.) alfalfa, red clover (*Trifolium pratense* L.), white clover (*Trifolium repens* L.), etc. Legumes are of great importance for the creation and development artificial grasslands and very valuable mulching surfaces that produce very high yields that are not only useful for silage, but also for drying.

Keywords: grasslands, legumes, Kosovo, distributions, nutritive values, diversity

1. Introduction

Forage crops and pastures provide the bedrock to sustainable agriculture. Grasslands cover about 70% of the world's agricultural area. They have a crucial role in terms of food production and in the delivery of ecosystem service such as water supplies, biodiversity and carbon sequestration. The grasslands of the world face a range of challenges from climate change including the effects of elevated atmospheric carbon dioxide, increasing temperatures, changes in precipitation regime and higher concentrations of ground level ozone [1]. Forage are defined as the edible parts of plants, other than separated grain, that provide feed for grazing animals or that can be harvested for feeding [2]. Forage plants provide humans with valuable ecosystem services, for example, they feed an estimated 1.5 billion cattle, 1.2 billion sheep, 1 billion goats and 0.2 billion buffalo around the world—supplying meat, milk and 33 other commodities [3]. Forages play a critical role of herbivores in nutrition, and are the foundation of most livestock rations. Nutritional requirements vary among types and classes of grazing animals; thus, what constitutes "high quality" forage for one animal may be "low quality" forage for another. For example, a dry cow will not require the same quality forage as a lactating cow [4].

IntechOpen

Livestock production can convert non-edible crops such as the forages into human food, with sustainable intensification possible when the system inputs and outputs are balanced [5]. Forage from grasslands is used to feed livestock and globally represent 26% of the land area, and 70% of agricultural area [6]. Forage crops consist in front of usually in grasses (*Poaceae*) or herbaceous legumes (*Fabaceae*) and some other botanical families. Bulky feeds are also termed forage and are produced from grass, cereal and legume cropping as described above, such as alfalfa, *Lolium* spp. or a mixture of the two. This forage can be provided to animals directly through grazing pasture land or in a processed form, such as hay (where water content is >15%) or dried (pelleted) biomass [7]. Due to the relatively recent forage cropping compared to other agricultural plant species, there were few improvements before 1900. Recently, agricultural trends and the global economic importance of forages, lead to cultivars have been bred by bred cultivars. These improvements are helped by many closely related wild populations which are potential for developing of new lines [8]. Most desirable improvements are increasing dry matter yield, crop durability and diseases resistance to particularly by pathogenic fungus and nematodes, dry matter digestibility, and nutritional content of this tissue. Arguably the greatest improvements have been made in breeding of *Medicago* spp., *Trifolium* spp., *Lolium* spp. and *Festuca* spp. [7].

1.1 Importance of forage and pastures in Kosovo

Actually Kosovo has around 1.9 million resident inhabitants, which mostly lives and works in urban areas. Last 20 years the Kosovar's society faced dramatically changes, lot of rural areas are abounded and consequently the arable lands and animal farms also, but these areas are not use as meadows neither as pasture now.

Based on actual data, the meadows and pastures cover 224410.83 ha of total surface of the Kosovo; moreover there, just a particular of these surfaces are distinguished clearly as meadows or pastures. Last agriculture census found that 65099.00 ha are meadows, 11724.27 ha pastures and 147587.56 ha are not clearly defined [9]. But these surface are used in alternate manner depends to farmers need. Trends of the immigration within or outside the country are quite high and lot of smallholders abandoned agriculture and farming. The number of small ruminants is decreasing year to year, which is affecting pastures disuse, and reduction of other ruminants has impact of the low use of meadows. Moreover, the cities extension and new infrastructure development is playing a huge impact on privation of agriculture land in general. In 30 years Kosovo has lost around 37000.00 ha of meadows and pastures. The largest areas with grasslands are in the mountains of Sharr, Bjeshkët e Nemuna, and Bjeshkët e Deçanit [9] (**Figure 1**).

Pastures are preferential to other animal feed resources, as having high nutritional value, and good taste, meeting nutritional requirements of herbivorous (cattle, sheep, goats, horses, etc.). Animals that feed on pastures have better health because they are daytime in the fresh air and other influence in health [10, 11]. Therefore, this also has an impact on the productivity and fertility of animals, are cheaper meals and there is no need to invest in food preparation, Plots in which grazing is continuously improved has an impact on the grazing quality. The fresh green forages dominate the humidity and this varies depending on the plant stage and ranges from 60 to 80%. Also, the dry matter obtained from new plants has high nutritional value as well as protein over 20%, and often brings up to 25% in dry matter. The amount of protein largely depends on the botanical composition where leguminous plants are dominant. Different factors affect forage crops, i.e., the botanical composition, plant botanical development, type and quality of land, climate, technology of grazing utilization (melioration, fertilization, grazing, etc.).

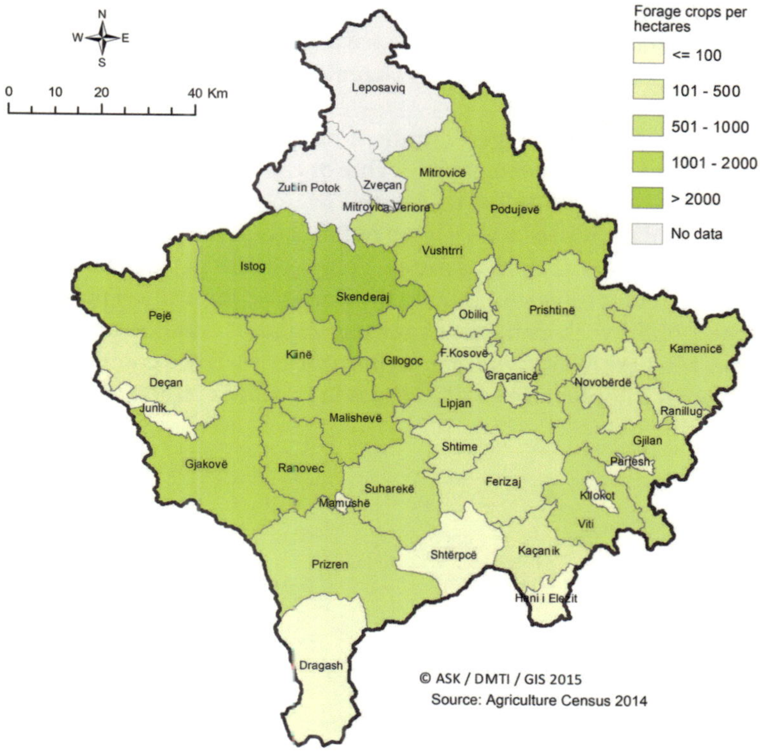

Figure 1.
Surfaces with forage crops in Kosovo [9].

Forages play basic role in Kosovo's beef cattle and milk industry while also enhancing crop diversity, wildlife habitat, and soil ecosystem services.

1.2 Distribution of pastures and forage crops in Kosovo

Meadows (14%) and pastures (30.2%) occupy significant land parts in Kosovo and they are very important source of fodder for feeding the livestock and protecting soil from against erosion [10, 12, 13]. Plants from *Poaceae* family are the basic forage crops which are dominantly in Kosovo. This group of herbs is divided into long, medium, and low herbs. Medium and long seeds are best suited for mowing and silage or drying for sanitary preparations, while short herbs are very suitable for grassland. In this group of plants many years old herbs are of great economic value, because they represent basic green foods in meadows, and artificial meadows. Intensive meadows and pastures can be found on over 2000 types of herbs [14]. *Trifolium* spp. has different distribution according as administrative region of the Kosovo. Peja's region covers the largest surface with *Trifolium* spp., or 34% of total areas 2084.50 ha (**Figure 2**) followed by Prishtina and Mitrovica regions.

In some extensive pastures highlighted about 200 species per 1 ha. This kind of herbs has high energy value. Moreover legumes which are also key components which were the participation of this group the pastures, where the participation of this group significantly increases the nutritional value of pastures. Legumes greatly contribute in creating and developing artificial grasslands and higher valuable mulching surfaces that produce higher yields that can be used for silage by drying. The most important types of fodder plants in Kosovo are: alfalfa (*Medicago sativa* L.), red clover (*Trifolium pratense* L.), white clover (*Trifolium repens* L.), Italian ryegrass

(*Lolium multiflorum* L.), bird's-foot trefoil (*Lotus corniculatus* L.), red fescue (*Festuca rubra* L.), etc. Surface of 15010.77 ha in Kosovo is cultivated by alfalfa. This specie covers 40% in regions of Prishtina and Prizreni, followed by Peja and Mitrovica region (**Figure 3**).

The *Poaceae* is the fourth largest family of flowering plants after the *Orchidaceae*, the *Compositae* and the *Leguminosae* [15]. One of the reasons for growing species in mixtures is getting a high nutrient feed with a balanced nutrient compounds [16]. Even in the natural meadows, a very large number of perennial legumes have been added, which add to the value of pastures. Their number in pastures and meadows is nearby 100 different types. Analyzing the floristic composition pointed clearly mainly of mesophilic with elements of xerophilic plants conditioned by relief shape and soil type. Mixture meadows are present in all sides of the country, but the largest surface, of total surface 6689.22 ha, is in Prishtina region or 32% whereas

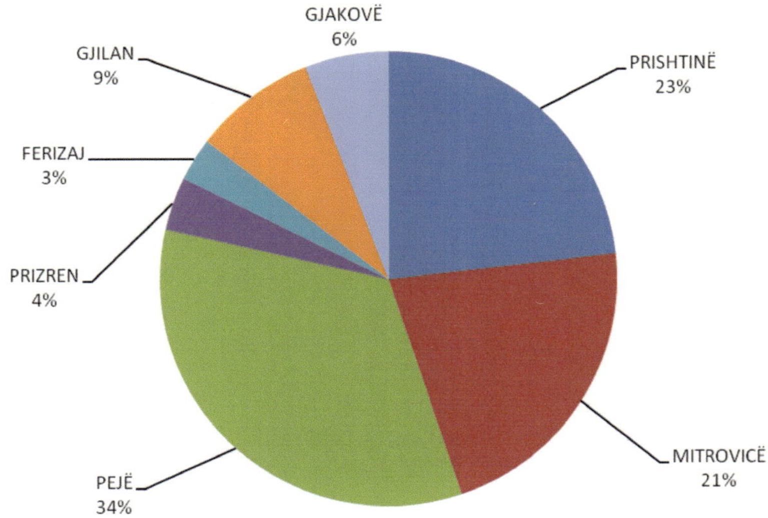

Figure 2.
Distribution of Trifolium *spp. by across Kosovo [9].*

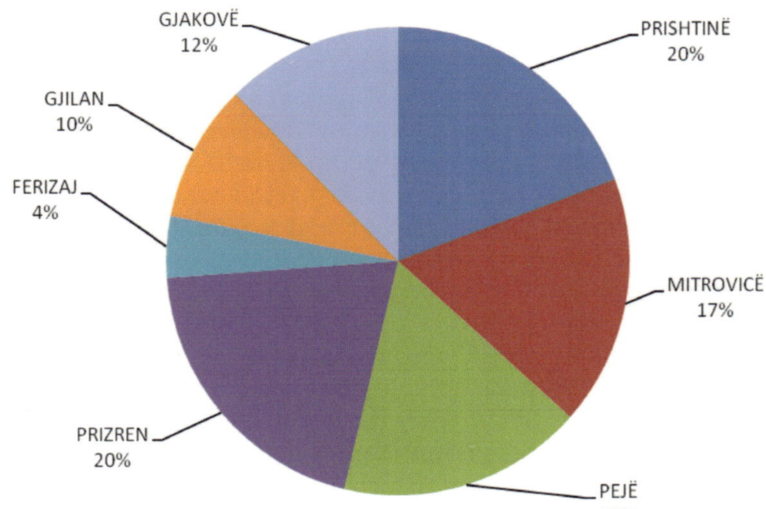

Figure 3.
Distribution of alfalfa by across Kosovo [9].

the Gjakova's region has the only 3% of surface with this forage. Distributions of the mixtures of meadows are presented in **Figure 4**.

In general, there is quality of the forage to the greater animal production response. While the concept of forage quality is fairly simple and straight forward, in reality it is rather complex. Furthermore, good management practice in grasslands provides high potential of carbon sequestration in soils, resulting in climate change mitigation.

1.3 Plant diversity of forage crops and pastures

Forage genetic resources play a very important role in food security and poverty alleviation, particularly in developing countries. Grasslands are among the largest biomes in world. Their area is estimated at and cover about 52.5 million square kilometers or 40.5% of the terrestrial area, excluding Greenland and Antarctica [2]. In the West Balkan (WB), the most important legume forage crops are alfalfa (*Medicago sativa* L.) and red clover (*Trifolium pratense* L.), which are mainly used as single N-fixation crops and rarely in mixture with grasses [17]. Alfalfa monoculture is still prevalent in Europe, where it is the backbone of organic crop-livestock systems, also significantly contributing to conventionally managed systems [18]. However, grass–legume mixtures are gaining new interest in Europe and elsewhere, owing to the energy and environmental costs associated with the synthesis and use of nitrogen fertilizers [17]. Sown pasture within arable rotations is important, mainly in temperate regions, for livestock production and fertility maintenance in rotations. Cut and carry fodder is locally important, especially for smallholders. Because mainly of the farmers in Kosovo store forage crops, i.e., Alfalfa for a longer period of time, to serve the animals as dry feed especially during the winter season.

1.4 Nutritive values of forage crops

Forage nutritive value partly depends on the available of nutrients in the plants for animals. Legumes also meet the requirement for greater self-sufficiency of feed proteins at the farm and the country levels [19]. Grass-legume mixtures are preferred over pure-grass forage stands throughout the world because they often increase the total yields of herbage and protein and offer balanced nutrition [20].

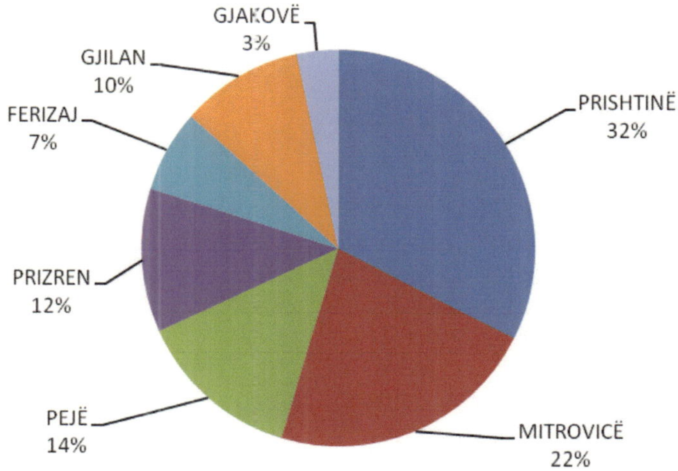

Figure 4.
The distribution of mixture meadows across Kosovo [9].

There are several grass species which could be successfully grown with legumes. Orchard grass (*Dactylis glomerata* L.) has been species of the favorite grasses for meadows in West Balkan (WB) due to its high annual forage yield and its ability to re-grow quickly after being cut or grazed. Tall fescue (*Festuca arundinacea* Sherb.), red fescue (*Festuca rubra* L.) and Italian ryegrass (*Lolium italicum* L.) are among the best forage grasses used in WB due to higher forage yield, fast re-growth and winter hardiness. In the point of view of yields (Dry Matter/ha), experiments at levels of Kosovo as well as the WB highlighted: (1) Association of Red Clover (RC) + Italian Grass (IG) was 12.0 t/ha DM; (2) Bird's-foot trefoil (BT) + Red Fescue (RF) 7.4 t/ha DM; (3) While the only Red Fescue (RF) provided 8.8 t DM [17]. As far as the nutritive values are concerned, combining RC + IG resulted in 11.38% Crude protein (CP) 33.90%, Crude Fiber (CF), 64.85% Neutral Detergent Fiber (NDF) and 38.19% Acid Detergent Fiber (ADF). The only Red Clover (RC) showed 10.69% CP, 33.32% CF, 72.63% NDF and 37.14% ADF [17]. The protein content (PC) is a quantitative trait and several studies have pointed out that there are a great number of genes involved in its control [21]. However, it is a challenge to manage the mineral supply of ruminants fed on grassland, because mineral concentrations in the herbage are influenced by a number of factors including species composition of the sward [22].

2. Conclusions

Forage genetic resources play a very important role in food security and poverty alleviation, particularly in developing countries. The distribution of plant species and their frequency differed from a location to another in Kosovo. Meadows (14%) and pastures (30.2%) are a significantly in Kosovo and appeared to be the key resources for feeding livestock and protecting the soil from erosion. Definitively forages play an important role in Kosovo's beef cattle and milk industry while also enhancing crop diversity, wildlife habitat, and soil ecosystem services.

Author details

Sali Aliu*, Imer Rusinovci, Shukri Fetahu and Dukagjin Zeka
University of Prishtina, Faculty of Agriculture, Prishtina, Kosovo

*Address all correspondence to: sali.aliu@uni-pr.edu

References

[1] Abberton MT, MacDuff JH, Marshall AH, Mike W. The genetic improvement of forage grasses and legumes to enhance adaptation of grasslands to climate change. FAO Report. 2008

[2] Allen VG, Batello C, Berretta EJ, Hodgson J, Kothmann M, Li X, et al. An international terminology for grazing lands and grazing animals. The Forage and Grazing Terminology Committee. 2011. DOI: 10.1111/j.1365-2494.2010.00780.x

[3] Hayduk D. The state of food and agriculture. FAO Report. 2016

[4] Castillo M, Forage Quality. Concepts and practices. Better Crops. 2017;**101**(4)

[5] Derner JD, Hunt L, Filho KE, Ritten J, Capper J, Han G. Livestock production systems. In: Briske DD, editor. Rangeland Systems: Processes, Management and Challenges. Cham: Springer; 2017. pp. 347-372. DOI: 10.1007/978-3-319-46709-2_10

[6] FAO. Challenges and Opportunities for Carbon Sequestration in Grassland Systems: A Technical Report on Grassland Management and Climate Mitigation. Rome: Food and Agriculture Organization of the United Nations; 2010

[7] Capstaff N, Miller A. Improving the yield and nutritional quality of forage crops. Frontiers in Plant Science. 2018;**9**:2018, 535

[8] Boller B, Green SL. Genetic resources. In: Boller B, Posselt U, Veronesi F, editors. Fodder Crops and Amenity Grasses. Handbook of Plant Breeding. New York, NY: Springer; 2010. pp. 13-38. DOI: 10.1007/978-1-4419-0760-8_2

[9] Zeka D, Gjokaj E. Kulturat e arave në Republikën e Kosovës, 2014. Prishtinë, Republika e Kosovës: Agjencia e Statistikave të Kosovës; 2016. 111p. ISBN: 978-9951-22-307-2

[10] Rusinovci I, Aliu S, Fetahu S, Bardhi N, Lombnaes P. The yield and mineral composition of some pure grass and their mixtures in agroecological conditions. In: Kosov 50 International Symposium on Agriculture, Croatia; 2016

[11] Ventura B, Marina A, Wittman H, Weary D. What difference does a visit make? Changes in animal welfare perceptions after interested citizens tour a dairy farm. PLoS. 2016. DOI: 10.1371/journal.pone.0154733

[12] Avdiu B, Aliu S, Fetahu S, Zeka D, Rusinovci I. The floristic composition of the natural pastures in massive of Novobërda. Agriculture and Forestry. 2018;**64**(4):235-241

[13] Rusinovci I, Aliu S, Fetahu S, Zeka D. The agronomic performance of some grass mixture in the Prishtina locality. Journal of Food, Agriculture and Environment. 2016;**14**(2):117-120

[14] Jerram R, Backshall J. Meadows and Enclosed Pasture (Chapter 7). UK: Natural England Publication; 2001

[15] Batello C, Mannetje L, Martinez A, Suttie J. Plant genetic resources of forage crops, pasture and rangelands. Thematic background study. FAO Report. 2008:5-7

[16] Thorvaldsson G, Björnsson H. The effects of weather on growth, crude protein and digestibility of some grass species in Iceland. Icel. Agr. Sci. 4, 19-36. under Ankara's ecological conditions. Turkish Journal of Agriculture and Forestry. 1990;**29**:69-74

[17] Cupina B, Vujic S, Krstic D, Djuric B, Aliu S, Manojlovic M, et al.

Performance of legume–grass mixtures in the West Balkan region. Acta Agriculturae Scandinavica Section B Soil and Plant Science. 2016. DOI: 10.1080/09064710.2016.120826

[18] Scialabba NE-H, Pacini C, Moller S. Food and Agriculture Organization of the United Nations (Smallholder ecologies). 2014

[19] Annichiarico P, Julier B, Louarn G, Maamouri A. Alfalfa intercropping and competitive ability. Legume Perspectives. 2014;**4**:29-30

[20] Albayrak S, Ekiz H. An investigation on the establishment of artificial pasture. Turkish Journal of Agriculture and Forestry. 2005;**29**(2005):69-74

[21] Aliu S, Rusinovci I, Fetahu S, Simeonovska E. Genetic diversity and correlation estimates for grain yield and quality traits in Kosovo local maize (*Zea mays* L.) populations. Acta Agriculturae Slovenica. 2012;**99**(2):121-128

[22] Kuusela E. Annual and seasonal changes in mineral contents (Ca, Mg, P, K and Na) of grazed clovergrass mixtures in organic farming. Agricultural and Food Science. 2006;**15**:23-34